About Island Press

Island Press is the only nonprofit organization in the United States whose principal purpose is the publication of books on environmental issues and natural resource management. We provide solutions-oriented information to professionals, public officials, business and community leaders, and concerned citizens who are shaping responses to environmental problems.

In 2000, Island Press celebrates its sixteenth anniversary as the leading provider of timely and practical books that take a multidisciplinary approach to critical environmental concerns. Our growing list of titles reflects our commitment to bringing the best of an expanding body of literature to the environmental community throughout North America and the world.

Support for Island Press is provided by The Jenifer Altman Foundation, The Bullitt Foundation, The Mary Flagler Cary Charitable Trust, The Nathan Cummings Foundation, The Geraldine R. Dodge Foundation, The Charles Engelhard Foundation, The Ford Foundation, The Vira I. Heinz Endowment, The W. Alton Jones Foundation, The John D. and Catherine T. MacArthur Foundation, The Andrew W. Mellon Foundation, The Charles Stewart Mott Foundation, The Curtis and Edith Munson Foundation, The National Fish and Wildlife Foundation, The National Science Foundation, The New-Land Foundation, The David and Lucile Packard Foundation, The Pew Charitable Trusts, The Surdna Foundation, The Winslow Foundation, and individual donors.

Earth Rising

Earth, isn't this what you want: rising up inside us invisibly once more?

—Rainer Maria Rilke, *Duino Elegies*

The future enters into us, in order to transform itself in us, long before it happens.

—Rainer Maria Rilke, *Letters to a Young Poet*

EARTH RISING

American Environmentalism in the 21st Century

Philip Shabecoff

ISLAND PRESS
Washington, D.C. • Covelo, California

Library of Congress Cataloging-in-Publication Data

Shabecoff, Philip.
Earth rising : American environmentalism in the 21st century / Philip Shabecoff.
p. cm.
Includes bibliographical references and index.
ISBN 1-55963-583-5 (cloth : alk. paper) — ISBN 1-55963-584-3 (paper : alk. paper)
1. Environmentalism—United States. I. Title.
GE197 .S43 2000
363.7'00973—dc21 99-050771

Printed on recycled, acid-free paper

Manufactured in the United States of America
10 9 8 7 6 5 4 3 2 1

To my grandchildren, Adam, Edward, Alexander, and Sophia,
and those who will inhabit the future with them

Contents

Preface xi

Acknowledgments xiii

CHAPTER 1
The Story until Now 1

CHAPTER 2
At the Turn of the Millennium 13

CHAPTER 3
Shades of Green: The State of the Movement 29

CHAPTER 4
Environment, Community, and Society 53

CHAPTER 5
The Business of America: Environmentalism and the Economy 83

CHAPTER 6
Playing Politics: Environmentalists and the Electoral Process 111

CHAPTER 7
Taming the Genie: Science, Technology, and Environmentalism 137

CHAPTER 8
Small World: America and the Global Environment 155

CHAPTER 9
Transforming the Future 177

Notes 195

Interviews 211

Bibliography 213

Index 217

Preface

This book is dedicated to my grandchildren and their contemporaries, who, if all goes well, will live through most of the 21st century. Their happiness and well-being will depend on many things, but nothing is more basic than the health of the physical world they will inhabit.

The environment—the land, the air, the water, and all that fills them—is the foundation of all life on earth. The welfare of individual humans and human societies depends ultimately on the solidity of that foundation. There is no subject of more crucial importance.

In the 20th century, as a result of human activity, deep cracks appeared in that foundation. Human numbers, economics, and technology began to put insupportable pressure on the systems that support life.

The 20th century also witnessed, however, the rise of modern environmentalism in response to the growing knowledge of the dangers into which we are thrusting ourselves. The rise of the environmental movement is, I have long believed, one of the most significant cultural developments of our time.

This book brings together what I have learned from more than two decades of thinking and writing about the environment and related subjects as a reporter for the *New York Times,* as publisher of *Greenwire,* the environmental news daily, and as the author of two books on environmental subjects. The earlier books were about the history of the American environmental movement and about the international effort to build an environmentally sustainable, socially equitable global economy. It seemed to me that an examination of the future of American environmentalism would be a useful contribution as we enter a new century and a new millennium.

Although my subject is environmentalism, I found when I had finished writing my first draft that I had also written what is, in essence, a critique of American society at the turn of the century. It is not possible, it seems, to discuss environmental problems and their solutions without addressing politics, economics, science, and social relationships. Those in the environmental movement, I concluded, will also have to involve themselves in the repair of these building blocks of our society far more deeply than they do today if they are to carry out their mission of preserving our physical habitat.

This book draws on a large number of interviews with environmentalists at the local, regional, and national levels; with scholars, politicians, and current and former civil servants; and with other activists who see environmental problems as issues of justice and community. Wherever direct or indirect quotations appear in the text without sources identified in endnotes, they are drawn from these interviews. I have conducted a partial review of the extensive and growing body of literature on contemporary environmental issues. The book is also based on my own sympathetic but detached observation and analysis of the movement. It reflects, therefore, both my own understanding and opinions and the understanding and opinions of a broad spectrum of thoughtful and informed people within and without the environmental movement. This is their book as well as mine.

What I have found suggests that in the waning years of the 20th century, the environmental movement, despite its great achievements, had not yet adequately prepared itself to meet the current and coming challenges, to transform a future that now seems to be filled with danger.

What follows is a close examination of the American environmental movement at the turn of the century: what it now is, where it stands in our society, how it is regarded by its friends and critics, what it is doing and not doing to meet the new challenges it faces both in the United States and globally, and, finally, what it must do if it is to continue to be a force for safeguarding the natural world and those who inhabit it in the next hundred years. Because some knowledge about the past order is needed to discuss the present and future, I begin with a brief summary of the movement's history.

Newton Center, Massachusetts
July 1999

Acknowledgments

Support, advice, and encouragement from many quarters made this book possible. My sincere thanks and gratitude go, first of all, to Joshua Reichert and The Pew Charitable Trusts and to Wade Greene and Rockefeller Financial Services for the generous financial support that enabled me to take the time to do the research and writing and to defray the many expenses associated with this extensive project.

My thanks go also to Island Press and its guiding spirit, Charles Savitt, not only for taking on publication of this book but also for serving as manager of my funding. Todd Baldwin, my editor at Island Press, was a full partner in this undertaking. At the outset, he provided trenchant ideas on how to proceed with the work and then gave me thoughtful, knowledgeable counsel in reshaping the manuscript. Pat Harris did a superb job of copyediting.

At my request, Josh Reichert also took the time to read the manuscript and to offer valuable suggestions. So did Stephen Viederman of the Jesse Smith Noyes Foundation. Susan Edelmann, former director of the liberal arts program at New York University's School of Continuing Education, gave the manuscript a sophisticated and meticulous reading. My wife, Alice Shabecoff, as she did for my previous books, not only carefully read the manuscript and offered much-needed advice but also assisted me in many other ways, including giving me a cram course on community economic development. The help of all these individuals has made this a better book than it otherwise would have been. Whatever errors of fact or judgment remain are mine alone.

I am very grateful to the nearly one hundred women and men—environmentalists, scientists, scholars, community and labor union activists, businesspeople, government officials and political professionals, journalists, foundation officers, and others whose names are listed at the back of this book—who submitted to long, sometimes arduous interviews. Their knowledge, experience, and, I hope, spirit, permeate this work.

My special thanks to John Adams, executive director of the Natural Resources Defense Council, who first suggested that I write about the future of the environmental movement, and to Peter Raven, director of the Missouri Botanical Garden, who provided provocative ideas and suggested sources for fleshing those ideas out.

Finally, I want to express my appreciation to Joan Martin-Brown of the World Bank, a pioneer of American and international environmentalism, whose wisdom, assistance, and friendship I have valued for many years.

Chapter 1
The Story until Now

About a century ago, in the middle of a thunderstorm high in the Sierra Nevada, a gaunt, bearded man climbed to the top of a wildly swaying evergreen tree, in order, he later explained, to enjoy riding the wind.

A few years later, the first head of the USDA Forest Service, a patrician, European-trained forester, was riding his horse through Rock Creek Park in Washington, D.C., when he had a sudden flash of insight. The health and vitality of the nation, he realized, depended on the health and vitality of the country's natural resources.

The White House, a few blocks away, was then occupied by a president who liked to shoot big game animals but who venerated wilderness and had conceived the idea, radical for the time, that the country's public forests, lands, and waters should be used for the benefit of all the American people, not just to increase the wealth of a few grasping robber barons.

These three charismatic, idiosyncratic contemporaries of a hundred years ago—John Muir, Gifford Pinchot, and Theodore Roosevelt—presided at the birth of one of the great cultural innovations of the 20th century: the modern environmental movement in the United States.

Over the course of the century, the movement grew and changed and achieved results far beyond the dreams of even those three visionaries. Many of their goals and dreams were realized as laws and institutions, in cleaner air and water, and in protected parks and forests, wildlife and wilderness areas. Over time, many, if not most, Americans, informed and prodded by the environmentalists, came to understand and integrate their values.

By century's end, however, it was an open question whether the environmental movement had reached the limits of its effectiveness. The problems had become much bigger, more complex and intractable, the solutions less clear. The forces arrayed against the environmentalists were stronger and more aggressive and sophisticated. The movement as a whole seemed increasingly subdued, less sure of its goals and how to accomplish them.

The roots of American environmentalism were planted well before the 19th century and are deeply embedded in the nation's history. Since colonial times, there were those who perceived—and some who lamented—the dramatic transformation of a pristine continent as a result of European migration, European technology, European economics, and European values. The great sweep of settlement across North America and the powerful tools and voracious demand for resources created by the industrial revolution profoundly changed the land and its people and did so with astonishing speed.

Some 150 years ago, transcendentalist Henry David Thoreau, sitting in his tiny cabin on Walden Pond in Concord, Massachusetts, was already mourning the loss of the wilderness and the debilitating effect of industrialism on the human spirit. In 1864, another New Englander, George Perkins Marsh, warned in his great work *Man and Nature; or, Physical Geography as Modified by Human Action*[1] that human activity could permanently damage the earth. The protection of Yosemite Valley by the state of California in that same year and the creation of Yellowstone National Park by Congress in 1872 were signals that the nation recognized the loss of its natural heritage and the need to preserve some portion of it for future generations. Across the young country, citizens in a few communities, troubled by the effects of sewage, unbreathable air, mining waste, or loss of forests and watersheds, organized locally to try to protect their surroundings and their health.

But American environmentalism—or conservation, to give it its birth name—was essentially a child of the 20th century, and Muir, Pinchot, and Roosevelt were indispensable in its creation.

For Pinchot and Roosevelt, conservation was a merger of science and democracy. Public lands and resources, they insisted, should be scientifically managed so that they would continue to serve the needs of all Americans, including future generations. Both men were sensible of the aesthetic and spiritual values of nature. But they preached what historian Samuel P. Hays called "the gospel of efficiency," which valued nature for its contribution to the public weal rather than for its beauty and other intrinsic qualities. They helped place huge areas of the public domain under permanent federal protection in the National Forest System and the National Wildlife Refuge System. Their efforts assured the environment of a permanent place on the nation's political agenda.

John Muir, a naturalist, writer, and mystic, introduced a different theme into the opening chapter of modern environmentalism. Imbued with a transcendental reverence for nature, Muir eloquently and passionately insisted that the natural world be preserved for its own sake as well as for humanity's. Everything in the universe, he maintained, is "hitched" to everything else, and humans tampering with any one part were interfering with the great cosmic plan. Muir also was a founder of the Sierra Club, the first of the major citizens' organizations that would increasingly rally to the defense of the environment over the course of the century. Among the other private conservation organizations founded in the first half of the 20th century and still in operation are the National Audubon Society, the National Parks and Conservation Association, the Izaak Walton League of America, The Wilderness Society, the National Wildlife Federation, Ducks Unlimited, and Defenders of Wildlife.

The early conservation movement, the "first wave" of environmentalism, was somewhat elitist. Its cadre and adherents tended to be affluent white Protestant males eager to protect wildlife for hunting and fishing and to preserve open space for aesthetics and recreation. Several early national conservation groups, including the Sierra Club, were for a time largely social organizations that existed to provide outdoor excursions for their members.[2]

Over the course of the century, however, rapid demographic, economic, and industrial growth created increasingly difficult, dangerous, and challenging risks to the environment, risks that could not be addressed by the tools of traditional conservation. The disappearance of wildlife, the fouling of the country's waters, the darkening of its skies from pollution, the loss of soil from erosion—especially during the dust bowl years—the onset of urban sprawl and disappearing farmland, and the introduction of hazardous chemicals and other substances into the air, water, and land by new industrial and agricultural processes grew as nagging concerns in the national consciousness.

But for much of the century, environmental dangers remained at the periphery of the nation's affairs. Public health officials and a few social activists tried to do something about the effects of environmental degradation on communities and workers. These efforts, however, were regarded as something apart from the nascent environmental movement.[3] Preoccupied by two world wars and the hardships of the Great Depression, Americans paid little attention to the effects of a ballooning population and rapid industrial growth on the natural world and on themselves. In the years after World War II, citizens were engulfed in a rising tide of materialism and a careless optimism tempered only by the cold war and the threat of nuclear annihilation.

In those same postwar years, however, powerful new technologies and explosive economic expansion created environmental pressures that could not be ignored. The country's quickly growing automotive fleet, powered by high-combustion engines, spread across the land on the new Interstate Highway System, pouring pollution into the air. Smoke from coal-fired power plants thickened the witches' brew of contaminants in the air, helping to produce a noxious haze that darkened the air in cities and dropped acid rain on streams, lakes, and forests. Nuclear testing put dangerous amounts of radioactive materials, including strontium 90, into the atmosphere. Off the coast of California, oil from offshore wells fouled waters and beaches as, elsewhere, did massive spills from the new supertankers. A growing culture of consumption created mountains of solid waste and rivers of sewage. Poisonous chemicals were carelessly buried or left in leaking steel drums to contaminate underground water supplies; one river burst into flame. The production of hazardous substances made workplaces dangerous, even deadly. The country's crops were drenched in insecticides, herbicides, and fertilizers, its livestock infused with synthetic chemicals and hormones.

Americans grew increasingly uneasy about the squandering of the country's once seemingly limitless resources, about the sullying of the landscape by industrial detritus and consumer trash. There was a growing suspicion that something was amiss in our affluent society, that we were fouling our own nest and poisoning our own wells. Our very affluence prompted many Americans to see environmental degradation as an obstacle to their search for a higher standard of living.[4] A growing body of scientific testimony seemed to verify that something was going very wrong. Neo-Malthusians, including Fairfield Osborn, Garrett Hardin, and Paul Ehrlich, warned that human numbers and consumption were outstripping what the earth could provide in perpetuity. Aldo Leopold's *Sand County Almanac*, an amalgam of science and ethics that is now one of the sacred texts of American environmentalism, admonished that humans are no more and no less than members of the entire community of life. Leopold called for a new land ethic that "changes the role of *Homo sapiens* from conqueror of the land-community to plain member and citizen of it."[5] Barry Commoner, the biologist, author, and political activist, and others pointed out that our technologies were breaking the chemical and biological cycles that sustain the planet. And in her acclaimed book *Silent Spring*, Rachel Carson presented clear, chilling evidence that the destructive technologies deployed by industrial society threatened all life, including human life.[6]

Public concern about the decline of the environment became a flood that could not be contained. It burst over the dam on April 22, 1970—the first Earth Day. Millions of Americans took to the streets and campuses to demonstrate their deep concern and to demand that environmental problems be addressed. On that day, environmentalism emerged for the first time on the national stage as an unmistakable mass social movement. The inchoate fears, anger, and longing of the public had been vivified into a suddenly potent political and economic force.

The immediate effect of the new tide of public opinion was to prod the federal government into action. President Richard Nixon, no "green" radical but keenly attuned to the political zeitgeist, stated that the 1970s "absolutely must be the years when America pays its debt to the past by reclaiming the purity of its air, its water and our living environment. It is literally now or never."[7] By executive order, Nixon created the Environmental Protection Agency, which became the single most effective federal tool for reducing pollution by corporations and municipalities—and for doing the research and education needed to alert the American people about threats to their land, air, water, and health.

Congress responded with a furious burst of bipartisan legislative activism, producing a spate of environmental statutes, from the National Environmental Policy Act of 1969, the Occupational Safety and Health Act of 1970, the Clean Air Act of 1970, and the Federal Water Pollution Control Act Amendments of 1972 to the Alaska National Interest Lands Conservation Act in 1980 and many more in between. In its totality, the explosion of congressional activism that produced these landmark environmental statutes must be considered one of the great legislative achievements in the nation's history.

State and municipal governments, in part because of new regulatory responsibilities passed on to them by Washington, also responded to the rising environmental impulse, creating their own environmental protection agencies and taking initiatives to address problems such as solid waste, compromised drinking water, and the need to provide open space for their citizens.

The new environmentalism emerged out of the social ferment and activism of the 1960s. It was an era of movements, notably the antiwar, civil rights, and feminist movements. Many of the senior cadre of today's major environmental organizations came from the militant campuses of that period. They believed that social change and political activism were the keys to protecting and restoring the environment. Unlike the older conservation groups, their focus was not on land and wildlife preservation but on pollution and toxic substances in the environment and their effects on human health. Out of this social activism sprang new environmental groups, including the Environmental Defense Fund and the Natural Resources Defense Council, whose chief tools were litigation and, later, lobbying for legislation designed to protect the environment. They took their battles to the courts to try to enforce the new environmental laws and to defend citizens threatened by environmental degradation. Other new groups, including Greenpeace, Environmental Action, and Friends of the Earth, used direct action and public information campaigns to alert Americans to what was being done to the natural world that sustained them.

A number of the national organizations pooled resources to form the League of Conservation Voters in 1970. The league monitored the environmental records of members of Congress and the executive branch and endorsed environmentally minded candidates on a bipartisan basis. Older conservation groups such as the National Wildlife Federation and the National Audubon Society broadened their agendas to take on the new issues of pollution, sprawl, and landscape degradation. The ranks of the environmentalists were reinforced by activists in the scientific community through the Union of Concerned Scientists and Physicians for Social Responsibility.

This period is often described as the "second wave" of environmentalism. Some commentators have called it the "golden age" of environmentalism in the United States, not only because of its legislative and political gains but also because it was a time when environmental quality also became an issue of democracy. Citizens across the country became aware of what was happening to their physical surroundings. Equally important, they also acquired a faith—not always requited—that in the American democracy, change was possible, that they could act as individuals and communities to obtain relief from the environmental dangers with which they were threatened. A watershed event of grass-roots activism captured national attention in 1978 when citizens of Love Canal, a neighborhood in New York, led by a courageous and crafty young housewife named Lois Gibbs, forced the federal government to pay for their evacuation from their houses, which had been built on the site of a toxic waste dump. The Love Canal victory inspirited communities around the country to address their local environmental concerns and drew many new recruits into the growing army of citizen activists.

The golden age of environmentalism, if such it was, came to an abrupt end in 1980 when Ronald Reagan entered the White House. Reagan was a simple man with a simple idea: government had become an unacceptably heavy burden to market capitalism and to the individual freedom of Americans. His goal was to get government off the backs of the people. In practice, this generally meant easing or removing regulatory requirements, particularly environmental regulatory requirements, on industry. He appointed environmental officials, notably Secretary of the Interior James G. Watt and Environmental Protection Agency administrator Anne (Burford) Gorsuch, who devoted their energies to reducing the ability of their agencies to protect natural resources and public health.

At the same time, corporate America, which had been caught off guard by the militant environmentalism that emerged in the 1960s and 1970s, began to mount an effective resistance. The business community, which had originally viewed anti-pollution efforts as a temporary if annoying fad, began to employ sophisticated skills, similar to those it had developed in its successful counterattack on the trade union movement, to fight environmental regulation and to counter the warnings and accusations leveled against it by the environmentalists.

In large part because the American people, alerted by the environmentalists through the news media, were paying attention to what was happening in Washington, President Reagan and his administration were unable to roll back the gains made over the course of the century. In fact, the membership rolls and treasuries of the environmental groups swelled to unheard-of levels as Americans demonstrated their concern by joining in record numbers. Gorsuch and Watt were forced to resign. And George Bush, Reagan's vice president, pledged to be the "environmental president" during his successful first run for the White House. A decade later, the environmental community was again able to rally enough public support to stalemate a ferocious assault on the environmental laws by right-wing radicals who controlled Congress.

In response to the counterattack against them, a number of environmental organizations, chiefly those operating at the national level, sought to develop new skills and tactics. They developed expertise in economic and political analysis, adopted more aggressive media and public outreach strategies, paid at least lip service to the disproportionate ecological ills heaped on the nation's poor and minorities, and looked for ways to achieve their goals without the assistance of the now less-than-sympathetic Congress and courts. Instead of attacking industry's every wrong environmental turn, some environmentalists sought negotiated settlements to pollution problems. The advocacy of market forces—as opposed to command-and-control regulation—as a tool for protecting the environment was a central feature of the new environmentalism.

These tactics were labeled the "third wave" of environmentalism. The most famous (or notorious, depending on one's perspective) example of this approach was the system of tradable air pollution permits proposed by the Environmental Defense Fund and accepted by President Bush and Congress for inclusion in the Clean Air Act Amendments of 1990. Proponents of the third wave called it a response to end-of-century political and economic realities, but it caused sharp divisions within the movement as critics complained that an excess of pragmatism was compromising essential goals. Although the market tools and accommodation tactics of the third wave may have blunted the counterrevolution, these critics say, they caused forward progress in protecting the environment to slow to a painful crawl.

As the 20th century drew to a close, it was clear that environmentalism had wrought profound changes in American life—to its landscape, its institutions, and its people. Since that first Earth Day, well more than one hundred pieces of major federal legislation affecting the environment had become law. Every state and most major cities had some kind of environmental protection agency. Wary politicians and battle-scarred corporations grudgingly conceded that the environmental movement was here to stay and was a potent force to be reckoned with. And given the enormous growth of population and economic activity, of production, consumption, and the generation of waste and pollution in the post–World War II era, imagine what the physical condition of the country would be had it not been for the environmental revolution. That our air is somewhat clearer and more breathable, that our water is somewhat cleaner and more drinkable in many places, that we are not buried in garbage, that some abandoned toxic waste sites have been cleansed, that some of our wildlands have been preserved from development and some of our threatened wildlife has been protected constitute almost miraculous achievements in the face of the economic juggernaut.

Even more significant, perhaps, is that environmentalism has changed the way most Americans look at the world and the way we live our daily lives. Public opinion polls consistently show that a majority of Americans consider themselves to be environmentalists. Most of us now think of a healthy environment as a basic human right. As writer Mark Dowie noted, "American environmentalism grew to become many things—world view, life style, science; to a few religion; and, eventually, a complex political movement."[8] Joshua Reichert, director of environmental programs for The Pew Charitable Trusts, probably the biggest single contributor to environmental advocacy causes among foundations, called environmentalism "the most significant social movement in America." Political scientist Michael Kraft found that "the environment had become a core part of mainstream American values. . . . It was as close to a consensual issue as one usually finds in U.S. politics."[9]

In the 20th century, environmentalism provided an intellectual and ethical context that enabled Americans, and people in much of the rest of the world, to see the harm that human activity was inflicting on the natural world—and on their own bodies. It established a legal and institutional infrastructure to help them come to grips with these ills and enlisted an army of activists, governmental and nongovernmental, at the local, national, and international levels, to work on solutions. An esoteric enthusiasm for a small elite at the beginning of the century, environmentalism had been transformed into a planet-wide value by its end.

But as a new century unfolds, the environmental movement faces challenging, often frightening, new issues. Problems such as climate change, acid precipitation, disappearing species, vanishing forests, crashing populations of marine life, spreading deserts, loss of topsoil, inadequate drinking water supplies, and dwindling farmland and other open space are compromising the balance of biological systems on the planet, threatening our quality of life, and narrowing the options for the continuing evolution of life, including human life.

At the same time, environmentalists will have to adapt to a rapidly changing economic, political, and social context. The end of the cold war, reawakened religious and ethnic hatreds, globalization of the economy and unchecked growth and concentration of corporate power, continuing inequity in the distribution of the nation's and the planet's wealth, an expected doubling of the world population, the explosion of new information and communication technologies, and the shifting sands of domestic politics all suggest that the voluntary organizations and governmental agencies charged with safeguarding the environment will be required not only to adjust their agenda but also to rethink the very nature of their mission and the means by which they pursue it.

Given the gravity of the problems, if environmentalists and their cause do not prevail in the next few decades, our habitat, our quality of life, and our democratic institutions could erode to the point that they might take centuries to recover. If the world's climate begins to change rapidly and dramatically, if the landscape continues to acidify, if hazardous, gene-altering synthetic substances continue to enter our flesh, if our per capita supplies of food and freshwater continue to dwindle, if we continue to waste nature's bounty and extirpate our biological resources, if we continue to deploy destructive technologies without recognizing their ultimate effects, then our children and grandchildren will live on a hot, dry, hungry, unhealthy, unlovely, and dangerous planet. The harshness of existence in such a world would very likely lead to the disintegration of communal life and the curtailment of personal freedom and opportunity. Environmentalists in that world would be seeking not to protect and restore nature and safeguard human health but to find ways for our posterity to survive amid the wreckage.

It should not come to that. Environmentalism has the latent strength to put us on a course toward a safe and pleasant ecological future, a better, more rational way of living on earth.

First, however, environmentalists will have to learn how to use that strength more wisely and effectively. They will have to find ways to rekindle the transcendental flame lifted by John Muir but now only a spark in their workaday institutions, to recapture the excitement and exhilaration in their cause that Muir found atop his storm-lashed tree. They will have to again practice conservation of the environment as it was envisioned by Roosevelt and Pinchot: as a core value of progressive politics, as an issue of democracy, as a means of bringing science to bear on the creation of policy, and as a means of achieving economic and social equity for present and future generations.

There will have to be yet another wave of environmentalism, one that is broader, more sophisticated, visionary, and aggressive and massive enough to stand against the tide of human numbers and technology, of ignorance and greed and willfulness, that threatens to propel us into an age of physical, biological, and cultural decline.

Chapter 2

At the Turn of the Millennium

In every age, no doubt, people have been sure that they lived in extraordinary times. And, of course, they were right. But it is hard to imagine there being many centuries as eventful, wonderful, and horrible, as explosive, transforming, triumphant, and terrible, as the 20th century was.

The century witnessed two enormously destructive world wars, a global depression, the end of colonialism, the rise and fall of fascism and communism, men walking on the moon, the breaking of the genetic code, unspeakable atrocities, and unprecedented acts of sacrifice and compassion by members of the human community.

But what may be the most important happening of the past hundred years attracted few headlines and, in fact, went largely unnoticed by most of the world.

In the 20th century, *Homo sapiens* became a force as powerful as nature itself in determining the condition of all life on earth. As Jane Lubchenco, then president of the American Association for the Advancement of Science, noted, "We now live in a human-dominated world." The changes occurring in the natural and social worlds, she admonished an audience of scientists in 1997, "are so vast, so pervasive and so important that they require our immediate attention."[1]

The Human Impact

In many areas of existence, our species has improved its status in ways unimaginable in past ages. We have conquered many dreaded diseases and substantially extended the span during which most of us can expect to remain alive and healthy. We have eased our labor while dramatically expanding our production of food, clothing, and shelter along with a cornucopia of other goods and amenities. We can communicate instantaneously across the globe and cross continents in a few hours. We have penetrated the atom and outer space, and we have learned to manipulate and even create the basic building blocks of life itself. We now know enough to make informed guesses about how the universe was created—and how it will end.

But our mastery and manipulation of nature has thrust us into dangers unknown to past generations. We are discovering anew that the tree of knowledge bears perilous fruit. Our use and misuse of the material world is altering the very physical, chemical, and biological systems that sustain life on earth. As Peter Raven, biologist and director of the Missouri Botanical Garden, pointed out, "There is not a square centimeter anywhere on earth, whether it be in the middle of the Amazon basin or the center of the Greenland ice cap, that does not receive every minute some molecules of a substance made by human beings."[2] By our voracious consumption of resources, we are rapidly extirpating other species with which we share the planet and, in so doing, are altering and compromising the course of evolution. The living flesh of every human is a repository of molecules of substances, some of them carcinogenic or mutagenic, created by human activity. As the National Academy of Sciences warned, "We are conducting an uncontrolled experiment with the planet."[3] The results of this experiment, unless in the near future it is controlled with more care and caution than is now the case, are unlikely to prove benign to our posterity.

Much of the dilemma springs from the explosive swiftness in the growth of human numbers and human power in the 20th century, particularly in the last fifty years. A thousand years ago, the total human population was somewhere between 250 million and 350 million, not much more than at the start of the previous millennium. Over the centuries, human numbers grew relatively slowly.[4] Not until the early 1800s did the global population reach the 1 billion mark. In the 1930s, the population had just reached 2 billion;[5] just in my lifetime, the number has tripled to 6 billion. Although the spread of family planning, the extension of human rights and educational opportunities to women, and rising living standards in many parts of the world are bringing birthrates down, the sheer weight of human numbers has achieved enormous momentum. A "medium-fertility scenario" prepared by the United Nations, which assumes that global fertility will stabilize at slightly more than two children per woman, projects that the global population will reach 9.4 billion by 2050 and 10.8 billion by 2150. If fertility rates remain at 1990–1995 levels, world population would reach an unthinkable 296 billion people by 2150.[6]

Human activity is already beginning to strain the ability of this small, finite planet to sustain us. Eroding soils, disappearing forests, crashing fisheries, inadequate drinking water supplies, shrinking farmland, and overcrowded cities testify that particularly in the poorer countries, too many people are competing for too few resources, in effect consuming the seed corn essential for future sustenance. An overly full world helps create the conditions that breed disease, addiction, crime, and often murderous competition for and conflict over living space and resources.

Even more than our numbers, however, our frantic economic activity, using the tools provided by ferociously powerful technologies, is straining our physical and biological systems to the point that they are beginning to rub thin the complex tapestry of life on this planet. Although population has tripled since midcentury, the global economy has expanded nearly sixfold in that period. The current aggregate world production of goods and services is roughly $30 trillion per year, a level that is already beginning to touch some of the limits of our capacity to produce resources and dispose of wastes. If recent growth trends continue, economic activity will more than quadruple in the next fifty years.[7]

Most of this soaring growth and consumption now takes place in the richer countries of the world. The global economic system, though enormously productive, is also highly inequitable. A world in which the 400 richest people control as much money as the 2 billion poorest people is a world, as Peter Raven commented, "characterized by social injustice."[8] The United States, with one-twentieth of the world's population, accounts for about one-fifth of global production.[9] But billions of people in the developing countries would like to emulate their rich relations in the North, and if they can, they will build their economies and consume and pollute in ways that could quickly overwhelm the ability of the planet to sustain us all.

In the rich countries, the chief threat to the environment and to our future comes not so much from the numbers of people as from the amount we consume and throw away and the tools and processes we use to produce what we consume. The materials and energy that keep us well fed and comfortable, that give us spacious houses, powerful private automobiles, and all the other amenities of late-20th-century life, are already taking a toll on the health of our habitat and threaten irreparable damage in the future. Our massive combustion of fossil fuels, the nonrenewable residue of organic life that has been buried deep below the earth's surface for millions of years, is changing the planet's basic carbon, nitrogen, and sulfur cycles. Our search for these fuels has led us to tear down mountains and to foul huge stretches of our coasts, to dim the brightness of our skies and blacken the lungs of our miners. Our answer to the need to increase the production of food has been, so far, to drench our land with ever more chemicals. Since Rachel Carson warned of the use of DDT and other chemicals more than thirty years ago, the use of synthetic substances that can sicken or kill people and wildlife has increased threefold. Our industries have rendered swaths of the country unfit for habitation—poisoned and unlovely brownfields, areas rendered dangerous by industrial activity. Land that has not been destroyed by industrial activity is being chewed up by developers for ring cities, residential subdivisions, strip malls, and second-home communities. Our technologies have been used to industrialize warfare and genocide and have produced calamitous accidents such as those at Chernobyl and Bhopal.

The Response of Environmentalism

It is difficult to imagine the predicament we would be in today if the environmental movement had not arisen to challenge the demographic and industrial juggernaut of the past half century. Perhaps the dead fields, sick children, choking air, poisoned waters, crumbling cities, and declining economies of parts of eastern Europe, Africa, and Asia are cautionary examples of where we might otherwise be. Because environmentalism produced new laws and institutions and rallied the public to demand protection, we in the United States have been able to treat the grosser and more immediate symptoms of ecological decline. Although the air still is not healthy in many areas and exacerbates illnesses such as asthma, even in American cities with the worst pollution, such as Los Angeles and Atlanta, the air is breathable most of the time. In contrast, the air in Mexico City remains unspeakably dirty, laden with pollutants that include particles of human feces; in coal-mining areas of Poland, children are taken down into abandoned mine shafts so that they can breathe relatively uncontaminated air. In the United States, dying waters such as Lake Erie have been restored to life, but in the former Soviet Union, indifference to environmental insults has turned much of what was the Aral Sea into a desiccated wasteland.

The rise of environmentalism in the 20th century is likely to be remembered as one of the landmarks of human social development—a time when the effects of human activity on the natural world were not only broadly recognized but also acted on by peoples and governments. But as the 21st century begins, it is clear that efforts to preserve and protect the environment and human health and well-being from our own assaults are only a beginning. Although we have made progress in reducing air and water pollution, limiting the damage caused by hazardous substances, and keeping some of our land relatively pristine, the task of cleaning up even the most obvious threats to our habitat is far from complete. We are only now coming to understand the complexity of the task that confronts us. Biologist David Ehrenfeld noted, "Historically, we have focused on unignorable symptoms and we have treated those symptoms as if they could be evaluated and managed in isolation, rather than regarding them as indicators of a terrible interference with the healthy functioning of a very elaborate network of chemical and biological relationships." Or, as William Clark of Harvard University's John F. Kennedy School of Government remarked: "We have knocked off some of the relatively easy problems. Now we run into the hard parts."

The Hard Road Ahead

The problems are, indeed, becoming hard. We are discovering that quick fixes and short-term solutions often not only fail to solve the problems but also make them worse. Requiring tall smokestacks to carry away pollution from coal-fired power plants, for example, may have eased problems near the plants, but it was painfully learned that the stacks spread acid precipitation hundreds of miles downwind, wreaking destruction on freshwater life and forests. It was relatively easy to get the first 50 or 75 percent of pollution under control. Getting the rest out will be much more difficult and expensive. In any case, merely controlling pollution ultimately fails—the problem is just moved from one place to another.

Experiments at the Hubbard Brook Experimental Forest in New Hampshire conducted by biologist Gene Likens and his colleagues had for many years shown that air pollution from the burning of fossil fuels was acidifying freshwater and forest soils. By the mid-1990s, data from Hubbard Brook showed that the problem was worsening in some areas despite the reduction of such pollution nationally.[10]

Problems of local, regional, and even national environmental deterioration, however, are now dwarfed by the danger to global ecological systems. The Environmental Defense Fund, in a strategic plan prepared in 1997, acknowledged:

> The environmental problems that face us, we now know, affect the Earth in its entirety—its climate, its oceans, its land, the interwoven lives of its inhabitants. Today, whole ecosystems are being compromised locally and globally. . . . An historic threshold has been crossed. A shift has occurred in the balance of strength between nature and humankind. We have passed, almost without noticing it, from a world in which the overall stability of the Earth's environment could be taken for granted to a world in which major, often irreversible manmade alterations of the environment are under way.[11]

At the top of the environmental agenda for the beginning of the 21st century is the need to confront climate change—the warming of the earth due to emissions of carbon dioxide and other human-produced gases. Author Charles Little pointed out that "for the first time in geological history, the amount of oxygen in the atmosphere is declining." The predicted consequences of climate change are by now widely known: melting ice caps and glaciers and rising seas that flood coastal areas, river deltas, and low-lying islands; ecosystems shifting too quickly for many species to adapt; increasingly violent weather episodes causing widespread damage to human habitat and economic activity; heat and droughts along with the spread of disease. All of these will pose new threats to human health and well-being.

The continuing loss of the earth's forest cover is both a cause and an anticipated effect of climate change. Now, human activity is stripping forests from the land at unsustainable rates in many parts of the world. In just the first five years of the 1990s, the earth's forested area declined by more than 11 percent, with most of the loss in the tropical countries.[12]

Deforestation contributes to another of the great environmental issues confronting the human community—the loss of species and the decline of the earth's biological diversity. As members of the Worldwatch Institute cautioned, "Arguably the single most direct measure of the planet's health is the status of its biological diversity—usually expressed as the vast complex of species that make up the living world."[13] Although some 1.7 million species of life-forms have been identified so far, no one knows the actual number of species existing on the planet. But at least 1,000 species are becoming extinct each year, and perhaps many more, almost all as the result of human activity. Harvard biologist Edward O. Wilson contended that the planet is undergoing a great "extinction spasm," the latest of several that have occurred over the millennia. The last such spasm took place at the end of the Cretaceous period, 65 million years ago, and wiped out the dinosaurs—probably, many scientists now believe, because of the effects of a giant meteor striking the earth. But the current spasm, Wilson asserted, "is human caused and can be human stopped."[14] "How many species can a system lose before it cannot compensate and collapses?" asked ethologist George Schaller. "The web of life that took millions of years to spin is being torn apart in a few hundred."[15]

Marine species are among the most threatened. As a result of overfishing, pollution, dumping of toxic wastes into the ocean, and climate change, many fisheries are in danger of crashing and fishermen in many areas are losing their livelihood. Catches of Atlantic cod dropped by 69 percent between 1968, at their peak, and 1992. Western Atlantic bluefin tuna stocks declined by more than 80 percent between 1970 and 1993. Many other fish populations are on their "last gasp."[16]

A world that will have to feed some 10 billion people in the 21st century cannot afford to lose so vital a part of its food supply. But fish stocks are not the only food source that is threatened. The world's per capita area of land in grain production has been declining steadily in recent years, in part because of soil erosion but chiefly because the world population continues to grow, yet almost all arable land is already under cultivation. A new technological fix of the kind that produced a dramatic increase in agricultural production during the "green revolution" of the 1960s and 1970s is highly unlikely. The late Henry W. Kendall, Nobel Prize–winning physicist at the Massachusetts Institute of Technology and chairman of the Union of Concerned Scientists, and David Pimentel, professor of ecology and agricultural science at Cornell University, found that "the human race now appears to be getting close to the limits of global food productive capacity based on present technologies." Although a doubling of food production by 2050 is achievable in principle, "the elements to accomplish it are not now in place or on the way. . . . A major reordering of world priorities is thus a prerequisite for meeting the problems that we now face."[17]

Other urgent, large-scale ecological problems will have to be addressed in the coming years:

- Acidification of the landscape as a result of combustion and dispersion of fossil fuels and other pollutants continues in the industrial world, including, despite measures taken to reduce the pollution, the United States and elsewhere. A transition from a system of industrial technology based on the burning of oil and coal to a system based on use of efficient, renewable sources of energy that do not threaten the environment and human health is fundamental. It will be hard to achieve, however, in the face of the fierce resistance by a powerful fossil fuel industry and its satraps.
- Although the global community has addressed the destruction of the earth's protective stratospheric ozone layer by industrial gases, more needs to be done, particularly by the developing countries.
- Massive industrial and agricultural use of toxic substances, including

many that can cause cancer or disrupt endocrine function, continues, and unknown thousands of toxic and radioactive waste sites have still to be cleansed.

- Freshwater supplies are already being stretched to the limit, and ways must be found to slake the thirst of the billions of additional humans as yet unborn. Already, about a quarter of the global population lacks access to clean drinking water.[18]
- The air inside our homes, offices, and factories and even many of our schools remains unhealthy and untreated.
- Farmland and other open space, particularly near population centers, is disappearing rapidly in the face of anarchic urban, suburban, and exurban sprawl and the development of rural farms and woodland for other purposes, such as second home developments and shopping malls.
- Wetlands, which offer a variety of indispensable ecological services, from flood control to spawning grounds for marine life, continue to be dredged and filled at a great rate.
- Significant new issues, such as the spread of genetically engineered crops and other products, and the potentially serious consequences of introducing massive amounts of antibiotics and hormones into livestock, appear with regularity

This is a daunting agenda for the environmentalists. And not only will they have to meet these challenges to the integrity of the environment at every level, they will also have to find ways to do so that enable and enhance a sustainable economy, reduce or eliminate the poverty that grips one-fifth of the global population, and ensure that there will be food, clothing, shelter, jobs, and a decent quality of life for twice as many people as inhabit the earth today. David Korten, author and president of The People-Centered Development Forum, warned that the United States and the world will have to confront "three interlocking crises: dehumanizing poverty, collapsing ecological systems and deeply stressed social structures."[19] Failure to address these crises, he warned, "will turn the 21st century into a global nightmare."[20]

Obstacles and Opportunities

The environmentalists' task is made far more difficult by the reality that the economic, political, and social contexts for environmentalism are dramatically different at the beginning of the 21st century from what they were on the first Earth Day, in 1970. These shifts are examined in detail in later chapters, but a brief overview will help illustrate the difficulties—and opportunities—facing environmental activists in the coming years.

A large sector of the American people and the politicians they elect to represent them are far less enthusiastic today about showering money on environmental problems than they were thirty years ago. In the years of prosperity and optimism following World War II, most Americans believed that the nation could afford the cost of environmental protection and environmental amenities without troubling the economy. No longer. Lester Lave, professor of economics, engineering, and public policy at Carnegie Mellon University, noted, "We are in an era when Congress and the public have on green eye shades and want to know the cost of protecting environment and how to do better." The Enterprise for the Environment, a task force chaired by former administrator of the Environmental Protection Agency (EPA) William D. Ruckelshaus and composed of representatives from industry, environmental groups, government agencies, think tanks, and academia, found that "in the next century the United States must (1) effectively address the remaining and newly emerging environmental issues facing the nation and (2) make the environmental protection system as efficient and low cost as possible."[21] As biologist, sage, and gadfly Barry Commoner pointed out, the increasingly high cost of the current practice of trying to control pollution rather than eliminate it by changing the technology of production "has *created* a built-in antagonism between environmental quality and economic growth" (emphasis in original).[22]

Changes in the structure of our economic systems have also increased the difficulty of dealing with environmental problems. In recent decades, the rise of the global corporation and rapid concentration of corporate power have created formidable impediments to the ability of national governments to control corporate behavior and enforce environmental laws and regulations. A small but growing number of corporations are embedding "green" principles into their business philosophy and operations. But as J. Clarence (Terry) Davies of Resources for the Future, former assistant administrator of the EPA, commented: "The business of business is business. That remains more true now than ever. To the extent that environmental initiatives don't pay and pay well, they are not likely to be embraced by corporations." Businesses, and particularly the trade organizations that represent them, have become increasingly energetic, sophisticated, and effective in opposing the environmentalists and turning back their initiatives. Through lobbying, advertising and public relations, political contributions, and "greenwashing" campaigns, the corporations have managed to stalemate, and in some cases roll back, progress in curtailing their environmentally damaging activities.

Environmental politics has changed even more dramatically in recent years. The first Earth Day took place in a political climate still largely dominated by the progressive tide that had been flowing since Franklin D. Roosevelt's New Deal in the 1930s. Americans still looked to government, especially the federal government, to be an engine of basic social change. In the 1960s in particular, enthusiasm for civic activism on issues such as civil rights, women's rights, and the anti-war movement demonstrated a broad faith that in our democracy, citizen action could help eliminate social evils. In Washington, D.C., a substantial degree of bipartisanship produced the landmark environmental statutes of the 1970s. The electorate sent to Congress a good number of principled, intelligent men and women who were dedicated to the public good. The Environmental Protection Agency attracted talented public servants who enthusiastically set about administering and enforcing the new environmental laws.

But New Deal progressivism, ebbing since the late 1960s, effectively ended with the election of Ronald Reagan to the presidency in 1980. Reagan and his conservative supporters acted on the belief that government was the problem, not the solution. Reagan sought to roll back the regulatory structure erected to control pollution, protect public and occupational health, and expand and conserve public lands. His administration's antagonism to environmentalism drove many able civil servants from the EPA and other federal agencies and caused others to hunker down and avoid actions that would cause displeasure in the White House. Much of the responsibility for environmental enforcement was transferred to the states, raising again the possibility that state governments would compete to attract industry by creating pollution havens. Faced with public support for environmental protection,

Reagan and his ideological allies were unable to gut the environmental laws or destroy environmental institutions. But they succeeded in abruptly halting the momentum of environmental progress that had been building for a generation.[23]

There also has been a substantial turn for the worse in the tone and content of political life in the United States. The bipartisanship and comity that marked the earlier environmental discourse in Congress and elsewhere in government has vanished. In its place is a no-holds-barred, win-at-any-cost politics emanating largely from the right wing of the Republican Party but infecting that party's entire program, which has poisoned the public policy dialog in Washington and throughout the country. Elections now are decided by negative advertisements and by a torrent of money, most of it from corporations, lobbyists, and wealthy individuals. Congress is dominated by representatives committed to an ideology of unfettered capitalism rather than to serving the welfare of all Americans. Like the know-nothing movement of the 19th century, this politics has tended to divide Americans, erode their trust in government and the democratic process, and alienate them from the political system.

The nation's environmental agenda has been a particular victim of this new politics. Because the environmentalists relied so heavily on governmental intervention to address pollution and resource problems, the declining faith in government and its institutions has blunted what were their most effective weapons.

The political climate of the 1990s also helped nourish a public backlash against environmentalism and environmentalists. Known as the "wise use" movement, this backlash was directed at environmental programs that restricted the use of private property or limited the exploitation of resources on which businesses and jobs depended. The most famous—or notorious—expression of this backlash was the outcry against efforts to protect the northern spotted owl and the ancient forests of the Pacific Northwest. This anti-environmental movement was provoked and financed to some extent by the energy, timber, and mining industries and other resource extraction industries seeking to reduce or eliminate environmental restrictions on their activities. Right-wing radicals, ultraconservative foundations, libertarian think tanks, and opportunistic demagogues enthusiastically helped stir the anti-environmental pot. But the backlash also represents genuine fear and anger among many Americans who believe, sometimes with reason, that environmental regulations will affect their property, their livelihoods, and their communities.

The anti-environmental movement has attracted only a small percentage of the American public. Public opinion polls consistently demonstrate that a substantial majority of Americans strongly support efforts to reduce pollution and protect land and resources. Nevertheless, the backlash is a fact of political and communal life with which the environmentalists need to deal seriously and with an understanding of the real problems faced by real people. By the end of the 1990s, they had done so with only limited understanding and limited success.

The geopolitics of environmentalism has also changed significantly in recent decades, in many ways for the better. Since the first Earth Day, a network of "green" activists at the international, national, and local levels has spread across virtually every country on earth. The United Nations conferences on the environment in Stockholm in 1972 and in Rio de Janeiro in 1992 helped inform the global populace about ecological dangers and moved environmental issues from the far periphery to near the center of the diplomatic agenda. Dozens of international environmental treaties have been signed, though not all have been ratified; some of them—such as the United Nations Convention on the Law of the Sea, the Montreal Protocol on Substances That Deplete the Ozone Layer, and the Convention on Biological Diversity—are of major significance in protecting the planet and its human and nonhuman inhabitants. Many governments now acknowledge that the ecological health of the planet is essential to national and collective security. Few countries are still without some kind of environmental protection agency. Most important, the global community is becoming well educated about the threats to its habitat. Environmental sociologist Riley E. Dunlap observed that recent years have seen "the emergence of widespread societal recognition of the fact that human activities are causing a deterioration of the quality of the environment *and* that environmental deterioration in turn has negative impacts on people."[24]

The end of the cold war, in theory, released international will and resources to focus on the pressing tasks of preserving the globe's life support systems. But the international effort to protect the shared environment has been negatively affected by a number of geopolitical trends. The reemergence of murderous ethnic, tribal, and religious conflicts in the wake of the cold war has deflected diplomatic energy and international resources from the great enterprise of achieving ecologically sustainable economic development for all. The Serbian drive to establish hegemony over the states that composed Yugoslavia, for example, diverted the attention and resources of the global community from the task of pursuing sustainable development and economic equity. The United Nations and other institutions of global governance remain weak and underfunded, few more so than the United Nations Environment Programme. The World Bank and other multilateral financial institutions, though no longer blind to the damage caused by single-minded industrial development and aware of the need for sustainable ecological systems, do not yet provide the incentives and disincentives needed to persuade developing countries and countries with transitional economies to do the right thing environmentally. Multinational trade organizations and agreements such as the World Trade Organization and the North American Free Trade Agreement appear to be giving commerce clear primacy over environmental goals and other values. Although governments generally give lip service to the urgent need to address the dangers to the earth's ecological integrity, few are willing to put their money or their laws where their official mouths are.[25]

Perhaps the greatest problem the environmental movement will have to face lies not in the external world but inside the heads of the American people. Although the great majority of Americans support environmental goals, that support may be shallow among many or most of them. Many people seem to have only a loose grasp of the dimensions of the problems and show little willingness to make any but the slightest changes in lifestyle. The very complexity of the emerging problems has a dampening effect on the public will to address them. Political scientist Walter Rosenbaum commented, "On the threshold of the twenty-first century, it is apparent that protecting the nation's environment is enormously more difficult, frustrating and costly than had ever been imagined."[26]

David Orr, professor of environmental studies at Oberlin College, contends: "The environmental crisis, the disordering of ecosystems, reflects a prior disordering of mind. This is in every possible way an intellectual crisis and mental crisis—and that makes it a crisis of education. . . . We have taught a generation to industrialize a planet without understanding the biosphere and ecosphere in which they are doing these things."

Whereas Orr and others believe that the intellectual failure to understand the environmental crisis stems largely from a failure in our educational system, other observers contend that the mass media also must bear a large share of the blame. With media ownership concentrated in fewer and fewer hands around the world and those owners more interested in power and profit than in an informed public, the focus of the press and the electronic media has grown increasingly frivolous. Although the public may support environmental protection, "the issues," pointed out Michael Kraft, "often fail to command the visibility needed to mobilize the public."[27] Lester Milbrath, an emeritus sociologist at the State University of New York at Buffalo, believes that "compared to the 1960s and 1970s, environmental awareness among the public is dropping." He traces much of the change to a media that, he said, "glorifies wealth and all the goodies produced by society and pays relatively little attention to the environment and to a long-term perspective." In the early 1980s, Milbrath wrote a book that described the environmentalists as the vanguard leading us into a new, green postindustrial society. Today, he is much more pessimistic. "The environmentalists are the vanguard, but they are not being listened to. They will be listened to only when our physical systems stop working. Life systems would have to go awry to the point where a couple of billion people die. That would wake people up. It is a grim scenario—so grim that people won't look at it."

In 1992, twenty-two years after the first Earth Day, political scientists William Ophuls and A. Stephen Boyan Jr. wrote: "Nature is still seen as either a mine or a dump and is treated accordingly. The basic laws of ecology are ignored, denied and flouted, and humanity continues to hasten down the path to ecological perdition." Although public attention drawn to environmental problems has promoted greater ecological awareness, "the watchword of industrial civilization continues to be what it has been since its inception: '*Apres nous le deluge.*'"[28]

This, then, is what confronts the environmentalists at the turn of the millennium. To preserve the physical world and its inhabitants, they have to find ways to alter profoundly the created world of economics, technology, industry, finance, and commerce; of politics, diplomacy, and ideology; of education, information, and communication. They will have to gain a better understanding of how people live and relate to one another in communities and how they think and learn and act, individually and as groups.

As eco-theologian Thomas Berry declared, "This is the great work of our age—to move the human situation from a destructive relationship with the Earth to a creative one."[29]

Is the American environmental movement up to the task?

Chapter 3
Shades of Green: The State of the Movement

Tim Hermach did not think much of the treaty drafted in Kyoto, Japan, in December 1997 to address the global warming problem. The treaty, which would require countries to reduce carbon dioxide emissions by a global average of 5 percent below 1990 levels, was no victory for the environment, he said. "If that is the best they can do, we are all dead." Hermach runs a small, take-no-prisoners organization called the Native Forest Council, which has as its goal the elimination of logging in the national forests. Zero cut. No trees killed. Period. He believes that the same principle needs to be applied to global warming: No industrial emissions of carbon dioxide. Zero. Period. By their willingness to compromise in Kyoto, he contended, the mainstream national environmental groups proved themselves wimps all too ready to acquiesce to further degradation of the earth's already damaged life support systems.

Fred Krupp, longtime executive director of the Environmental Defense Fund, one of the leading and most active national environmental organizations, saw the Kyoto treaty as a major step by the international community toward meeting the dangers of global warming. He believes that environmentalists were key actors in persuading the governments of the United States and other countries to agree to a pact. He is proud of the role that members of his staff played in pushing for the treaty and for the concept of using economic incentives to induce industry to reduce carbon emissions. Krupp and his organization have been in the forefront of "third wave" environmentalism, which employs market forces and negotiation with business and government, along with lobbying and litigation, as tools to achieve environmental goals.

William Clark of Harvard University's John F. Kennedy School of Government is "skeptical" that any of the traditional environmental groups have a "unique, crucial, or powerful role" in shaping what happens to the environment. He finds that some industrial organizations, such as the Electric Power Research Institute (EPRI), an arm of the utility industry, which produces much of the carbon emissions through the burning of fossil fuels, play at least as much a role in protecting the environment as do the green groups.

As a nurse, mother, and citizen, Terri Swearingen is concerned about global warming. But as a local environmental activist, she does not pay attention to that global issue. Her energies and intelligence are focused on one community and one problem: the dangers to health posed by a toxic waste incinerator in East Liverpool, Ohio.

The Movement Defined

What is the environmental movement? *Is* there an environmental movement? Some critics, including some environmentalists, believe that a coherent, organized environmental movement no longer exists—if ever it did. Barbara Dudley, former director of Greenpeace USA, contends that there has been no national environmental movement since the 1970s, "just a lot of different people organized around saving a forest here or blocking a toxic waste incinerator there. When people talk about a movement, they are talking about organizations. But there is no leadership of a movement." Stephen Viederman, president of the Jesse Smith Noyes Foundation, which funds many grass-roots projects, thinks that there is not one environmental movement but "three or four movements very different from each other" dedicated to different goals and employing different skills and tactics. Author and veteran environmentalist Charles Little said that there is a movement, but in recent years "it has become moribund because it has lost its moral foundation. The big organizations have hired suits to run them and have undertaken economic analysis instead of sticking to their ethical last. No one takes them seriously anymore."

My own view, based on more than two decades of close observation of the evolution of the movement, is that although environmentalism has become extraordinarily diverse and complex since the first Earth Day, with many parts, differing agendas, and a proliferation of institutions, and although its components are often in disarray and sometimes in conflict, it is a coherent social and cultural movement unified by a broadly shared—if diffuse—view of the world, the world's most fundamental problems, and the causes of those problems. It is also, as environmental sociologists Angela G. Mertig and Riley E. Dunlap have found, "one of the most successful contemporary movements in the United States and in Western Europe."[1]

Some deep fault lines separate sectors of the environmental movement, none of them deeper than the argument—specious in my view—over a human-centered versus a nature-centered environmentalism. But the many parts of the movement come together in a set of goals that can be loosely described as preserving and restoring the natural environment and protecting the human species—and other life—from degradation, depletion, and destruction of the natural environment by human activity. "Reduced to its essentials," wrote Walter Rosenbaum, "environmentalism springs from an attitude toward nature that assumes that humanity is part of the created order, ethically responsible for the preservation of the world's ecological integrity and ultimately vulnerable, as are all earth's other creatures, to the good or ill humans inflict on nature."[2] Scholar Lynton Keith Caldwell noted, "All goals of the environmental movement have not been defined, but among them the attainment of a sustainable economy of high environmental quality is a widely shared objective."[3] With Lester Milbrath and others, Caldwell considers the environmental movement to be leading the way to a postmodern society characterized by a new worldview dominated by a "planetary paradigm." The movement is perforce highly political, but it is neither left nor right, neither capitalist nor socialist. It is a progressive movement in that its proponents believe that the world—its people, its institutions, its technology, its economics, its politics, and, of course, its physical health—can be changed for the better. It is conservative in that it seeks to conserve the natural world and protect life from radical technologies and destructive economic activity.

Michael McCloskey, longtime chairman of the Sierra Club, who has been a close observer of the environmental movement for more than forty years, contended that it has become a "mature" social movement. That means, he said, that it is highly variegated, is broadly accepted, has a vested interest in maintaining existing systems for fighting pollution and defending public lands, and is associated with the prevailing political and economic establishment. It has also, he said, become divided into three camps: mainstream, radical, and conservative. Members of mainstream groups are basically "pragmatists" seeking incremental reforms and trying to work marginally to reduce pollution, conserve energy, and protect nature by making basic changes in the nature of the political and economic systems. The mainstream groups work with government and the political parties and "while often battling with business and industry, [don't] see them as our permanent enemies either," McCloskey explained. The mainstream includes his own Sierra Club as well as such groups as the Environmental Defense Fund, the Natural Resources Defense Council, the National Audubon Society, the National Wildlife Federation, The Wilderness Society, and the World Wildlife Fund. Greenpeace is on the boundary between the mainstream and the radical camp.

The radicals, among whom McCloskey numbers the militant Earth First! activists, deep ecologists, and many groups at the local level, particularly those fighting toxic and nuclear contamination and members of the environmental justice movement, are "alienated from establishments of all kinds," including the mainstream environmentalists, who, they think, "have sold them out." The radicals, he said, are anti-government as well as anti-business. They seek fundamental changes in the political and economic systems and give the need to protect nature primacy over the need to protect humans.

The conservatives include members of hunting, fishing, and land preservation groups such as Ducks Unlimited, The Nature Conservancy, and, McCloskey contended, the National Rifle Association. The right wing of the environmental movement is suspicious of government but rarely criticizes business. It tries to reach its goals largely through the private sector.

In recent years, he added, both the radical and the conservative wings of the movement have moved further to the extremes. And even the mainstream has been dividing into sometimes antagonistic camps, including what McCloskey described as the "stand patters," who will not give up any perceived environmental gains, and the accommodators, who will compromise for the sake of preventing further erosion. In the early 1990s, the mainstream groups split down the middle in a nasty exchange over whether to support or oppose the North American Free Trade Agreement. Several, among them the Environmental Defense Fund and the National Wildlife Federation, supported the treaty because they believed that it would lead to greater prosperity, particularly for Mexico, and this in turn would create conditions favorable to stronger environmental protection. Others, such as the Sierra Club and Friends of the Earth, were against the treaty, believing that it would let considerations of free trade and economic growth ride roughshod over environmental laws. Charges of selling out the environment flew back and forth among the groups. "The movement against itself is one of the big problems," McCloskey said.

Although McCloskey's analysis is pretty much on the mark, there are a number of groups that do not fit neatly into any of his categories—groups such as the U.S. Public Interest Research Group, founded by Ralph Nader, and Friends of the Earth, which employ aggressive tactics or seek changes in basic tax laws. Social ecologists, including members of the E. F. Schumacher Society, seek radical shifts in the economic structure of society, but to a conservative, locally oriented lifestyle and system of production. Organizations such as the Union of Concerned Scientists, whose membership virtually defines the word *establishment,* seek radical changes in existing economic and political governance to protect the world from disaster. A few organized or quasi-organized groups whose members identify themselves as environmentalists, such as the saboteurs who bombed a ski resort in Colorado to try to block its expansion, are by no stretch of the imagination part of the social movement known as environmentalism. As deep ecologist Bill Devall pointed out, "non-violence is a central norm for radical environmentalism."[4]

In recent years, the number of environmental organizations at the international, national, regional, state, and local levels has grown so rapidly and the structure of the movement has become so complex that it resists attempts to compartmentalize its components. An attempt, however, was made by Adam Werbach, who became president of the venerable Sierra Club at the tender age of twenty-three. Werbach divides the environmentalists into "druids," who defend nature for its spiritual qualities; "polar-fleecers," who want to preserve forests, streams, and natural areas for their recreational value; "apocalyptic" environmentalists, who warn about the destruction humans are bringing on themselves and the planet; "eco-opportunists," who engage in lobbying, litigation, and politics; and "eco-entrepreneurs," who recognize the economic problems of dwindling resources and the economic opportunities in protecting the environment. "I don't care why someone cares about the environment," Werbach wrote, "only that they do."[5]

The environmental movement is not only diverse; it is also a constantly shifting, evolving organism. Some formerly influential organizations such as the Environmental Policy Center have quietly exited the national stage. Some, including American Forests, which has been involved in the nation's woodland affairs for more than a century, have demonstrated great institutional resilience and staying power. Others, such as Greenpeace, have suffered from loss of membership and internal disarray and lost some of their clout. New groups, including the National Environmental Trust and the Environmental Working Group, continue to emerge and play significant roles. Many institutions have sprung up to deal with specific issues, among them Ozone Action, American Rivers, SeaWeb, the Rainforest Action Network, the Center for Marine Conservation, and the Rails-to-Trails Conservancy. Although the environmental justice movement is often at odds with the traditional groups, it has established a whole new network of institutions and enlisted thousands of new constituents who struggle to link environmental protection with human and civil rights. Regional groups such as Great Lakes United and the Greater Yellowstone Coalition take on issues that cross state borders. Ecotrust, founded by a former vice president of The Nature Conservancy, Spencer Beebe, helps citizens practice conservation-based economic development at the community level. Think tanks such as the World Resources Institute and the Worldwatch Institute provide research and policy tools on a wide range of global environmental and economic issues. The Orion Society helps enlist literary values in support of the environmental enterprise. Second Nature, founded by Anthony Cortese, former dean of environmental studies at Tufts University, is dedicated to fostering environmental education, particularly in higher education. In Congressman Sherwood Boehlert's district in upstate New York, there is an organization called Kids Against Pollution. Groups of local activists spring up almost daily to fight a particular factory, waste dump, or incinerator siting or to protect a piece of open space or body of water. It would take a thick book just to list all of the environmental groups in the United States at the end of the 20th century. In fact, the National Wildlife Federation publishes such a book, although its 500-plus pages are far from comprehensive.

How big is the environmental movement? It depends on how it is defined. If all Americans who described themselves as environmentalists were included, the movement would encompass two-thirds or more of the American people, according to many public opinion polls.[6] The dues-paying membership of the national organizations changes with the political and economic climate and with the presence or absence of significant environmental disasters. In the mid-1990s, the aggregate membership of the major groups was probably around 12 million to 14 million.[7] Some of those members were counted twice because they belonged to more than one group. Of course, membership in these groups does not necessarily imply activism. A majority of the members of most national groups probably are passive dues payers, who sometimes can be aroused to write to their representatives in government or to write letters to the editor of the local newspaper.

Grass-roots organizations are another matter. Their members tend to be intensely involved in their local issues because often their health, the health of their families, and their property are immediately at risk. There are no reliable estimates of the number of grass-roots organizations and their membership. The Center for Health, Environment and Justice (formerly the Citizens Clearinghouse for Hazardous Waste), run by Lois Gibbs, organizer of the Love Canal Citizens Committee, works with a network of some 8,000 to 9,000 local groups with memberships ranging from a half dozen activists to several hundred.[8] There are undoubtedly many more local groups with some focus on the environment. An educated guess at the aggregate membership of all environmental organizations in the United States in the late 1980s put the number at about 25 million.[9] Although the mainstream groups had their ups and downs in the 1990s, the number was probably significantly higher at the end of the decade.

Readiness for Battle

Early in 1995, I encountered the head of one of the major environmental groups during a meeting in Washington, D.C. The 104th Congress, with its "contract with America," had recently taken over Capitol Hill and was mounting a fierce assault on the environmental laws and the federal institutions that enforce them. I asked the environmentalist what his organization was doing to counter the anti-environmental tide. His answer: "Well, we've already issued a press release."

The mainstream, Washington-based environmentalists were not prepared for the unrestrained ideological and political counterattack on their agenda. Only two years previously, their hopes for achieving significant breakthroughs had soared with the election of Democratic president Bill Clinton and, especially, Vice President Al Gore, who had been the most committed environmental advocate in the U.S. Senate. But the new tenants in the White House, preoccupied by the economy and other matters such as the Whitewater and campaign financing investigations, did little regarding the environment. At the same time, the political strategy of the Republican minority in the 103d Congress was to deny legislative progress on the environment or any other issue for which the president and his party could claim credit. The strategy was largely effective. But it was only a foretaste of what would happen later when the Republicans, fully dominated by the conservative wing of their party, gained a majority in both houses and placed anti-environmentalists in many key committee chairs. All the anger and resentment conservatives felt over the progressive impulse that had dominated Congress for most of the years since Franklin Roosevelt's New

Deal seemed to explode after the Republicans took over Capitol Hill in 1995. And much of that resentment was directed against environmentalism and environmentalists.

Meanwhile, much of the American public assumed that the environmental issues were being satisfactorily addressed, and in the absence of any attention-grabbing environmental disasters like the *Exxon Valdez* spill, membership and funding began slipping backward for many of the national organizations. With limited resources and divided by their competition for money and members and by differences over agendas, tactics, and strategy, the demoralized environmentalists were initially no match for the furious offensive by the political right wing and its allies in the business community. Not only was the lobbying clout of the mainstream environmentalists greatly diminished, but also the movement clearly lacked sufficient ability in organizing and communication to rally the American people in defense of the framework of environmental protection that had been so painstakingly built over the previous quarter century. Press releases were a flimsy line of defense.

"If the American environmental community at Earth Day 1995 seemed bewildered and directionless, it was for good reason," noted political scientist Christopher J. Bosso. The new Republican majority seeking to roll back decades of progress in federal environmental protection, he said, was reinforced by well-organized property rights and "wise use" groups allied with industry, which were also seeking to weaken state laws. "An apparent lack of public concern about environmental issues and a general mood of fiscal retrenchment," he wrote, "encouraged those forces to pursue agendas whose audacity stunned even environmentalists who agreed on the need to overhaul existing command-and-control regulatory approaches."[10] After their initial bewilderment and lethargy, the national environmental groups got their act together sufficiently to rally public support and to stalemate the efforts to gut the environmental laws and agencies. New organizations, such as the National Environmental Trust and Environmental Media Services,[11] brought crucial organizing and media skills and energy to the effort to stem the anti-environmental tide. Overreaching and virulently partisan rhetoric by many of the conservatives also helped turn the American people against the excesses of the 104th Congress. The environment is not usually an issue of high political salience, but when the quality of their water and air and the health of their children are threatened, Americans can be roused to anger. The frontal attack on the environmental laws was blunted. Efforts to weaken environmental protection went underground in the form of riders attached to other pieces of legislation, a tactic partially but not completely countered by resistance from the White House. The worst did not happen. But several significant actions to weaken environmental protections, such as the so-called salvage-logging rider, which handed timber companies a key to many formerly protected areas of national forest, made it through Congress and avoided a presidential veto.

Foundations and the Environmental Movement

The role played by the National Environmental Trust in the campaign to preserve the environmental laws underscored the growing significance of the foundation world in not only supporting but also giving direction to the environmental movement. The trust, which started life in 1993 as the Environmental Information Center, was a creation of The Pew Charitable Trusts, one of the biggest foundations and probably the biggest contributor to environmental causes. Its director of environmental programs, Joshua Reichert, was concerned that the national green organizations did not have the capacity and institutional nimbleness to meet the new realities of environmental politics and economics. They were not, he believed, good at mobilizing people at the state or local level, to which much decision-making power was devolving; at building coalitions; or at making use of information and media tools to achieve their goals. Therefore, the Pew Trusts funded the creation of a new organization headed by Philip Clapp, an experienced former Senate aide, and staffed by others with the organizational and communication skills needed to combat the growing influence of corporate money and the political right.

Wade Greene, who manages anonymous grants by members of the Rockefeller family, said, "Many funders feel that the major environmental organizations have grown too comfortable as they have aged and are no longer politically effective."

Foundations have long supported environmental causes. The Natural Resources Defense Council was basically created with money from The Ford Foundation. The World Resources Institute was brought into existence and nurtured with funds from The John D. and Catherine T. MacArthur Foundation. The Environmental Grantmakers Association has nearly two hundred members, foundations that regularly contribute funds to environmental groups and causes. Foundations contribute about one-fifth of the income of the environmental organizations.[12] But The Pew Charitable Trusts and a number of other foundations have been taking a much more active role in the direction of environmental policy, not only creating institutions such as the National Environmental Trust but also staffing them, setting their agendas, and giving them their marching orders. Jonathan Lash, president of the World Resources Institute, noted: "There has been a change in the nature of foundations. They used to be a source of intellectual venture capital. Now they want to be players in shaping and running the program."

Reichert acknowledged that the Pew Trusts are having a "huge influence" on the current direction of the environmental movement. He said he knows that "a lot of people" in the movement are unhappy with the foundation's approach but added, "I am not apologetic." Although the Pew Trusts may spend millions on environmental issues, there is much more funding on the other side of those issues. So, he said, foundations need to be much more efficient in the way their money is used: "I want to be sure our investments produce measurable results."

The head of another foundation that is a major supporter of environmental causes believes, however, that "foundations are just as mixed a bag [as the environmental groups]. In fact, they are more so because few foundations have discipline and focus and they lack accountability," said the official, who declined to speak for attribution.

The Jesse Smith Noyes Foundation, whose chief executive, Stephen Viederman, is critical of the national environmental movement, makes most of its grants to environmental justice organizations and grassroots and community-based groups. A number of other foundations support local activism. But a large proportion of money from environmental grant makers goes to the big national and regional organizations. William Shutkin, former director of Alternatives for Community & Environment, based in a low-income area of Boston, said that the big foundations tend to ignore organizations such as his. They may have "a ton of money," he said, "but if you are not right at the core of the mainstream organizations, it's really hard to break into that circle."[13]

Of course, a number of conservative foundations, along with many corporations and trade associations, are major funders of the anti-environmental movement. Right-wing foundations such as the Charles G. Koch Charitable Foundation, the Scaife Foundations, the Adolph Coors Foundation, The Lynde and Harry Bradley Foundation, and the John M. Olin Foundation, Inc., pour a considerable amount of money into libertarian and uncompromising free-market organizations such as The Cato Institute, The Heritage Foundation, and the Competitive Enterprise Institute. "Wise use" and "property rights" groups hostile to environmentalists and their causes often get financial support from such sources.[14]

Currents in the Mainstream

Although it weathered the storm of the 104th Congress, the national environmental movement was obviously in a state of disrepair at the end of the 20th century, its flaws and failings laid bare for all the world to see. The movement, which had changed the face of the nation over a generation, badly needed to take a long look at itself and to regroup, reorganize, and prepare to move forward again. Progress in preserving the environment had slowed to a near halt, Fred Krupp of the Environmental Defense Fund said, "because the world is changing and we haven't adapted." And, in fact, a number of the big national groups did go through intensive soul-searching and reorientation.

A self-assessment by the Environmental Defense Fund (EDF) conducted in 1997 concluded that the group had been spreading itself too thin over too many issues. Future efforts, it decided, would concentrate on four areas: working to mitigate climate change, protecting the oceans, protecting human health from toxic chemicals and pollution, and defending and restoring biodiversity. EDF also adopted a new mission statement, which includes, among other things, a commitment to "economic and social systems that are equitable and just. We affirm our commitment to the environmental rights of the poor and people of color."[15] Krupp said that EDF is now fully committed to working at the local level. One of the biggest changes in recent years, he said, has been a move toward pluralism in governance: "We have to look at how power has become more diffuse at the community level." In 1998, Krupp reported, EDF was at an all-time high of 300,000 members and a budget of $26 million.

The Natural Resources Defense Council (NRDC) has been one of the premier environmental lobbying and litigating organizations since its founding in 1970. Its cadre of lawyers, scientists, and other professionals has focused on promoting specific pieces of legislation, such as the Clean Air Act, or on saving a particular species or forest. But John Adams, who has been executive director of the organization since it was founded, noted that in the mid-1990s, "we made a big shift from just litigation and lobbying. We are now doing campaigns. We are not just fighting to preserve sharks but conducting a campaign to protect the ocean in all of its aspects—pollution, overfishing, markets. That is a very significant change. Our people aren't isolated in specific pieces of litigation." All legislative activity by the organization is now coordinated by a single office. NRDC also increased its membership from 250,000 at the beginning of the 1990s to 450,000 in the latter part of the decade and, with an annual budget of about $25 million, put itself on a more financially secure footing by raising a multimillion-dollar endowment.

The National Audubon Society, one of the old-line conservation groups, also went through a period of critical self-examination after an internal dispute over mission and strategy that led to the resignation of half the board and a change in leadership. John Flicker, the aptly named president of the organization, said that Audubon, which had become engaged in a broad spectrum of environmental issues, decided to go back to its roots and concentrate on birds and other wildlife. A major lobbying force in Washington, D.C., for many years, Audubon decided to decentralize and use its resources chiefly to support its 525 chapters around the country. "Our members told us they wanted us to do this," Flicker explained. The head of another national group said that Audubon had essentially "dealt itself out of the Washington game," but Flicker contended that "our power in Washington comes essentially from our grass roots speaking back home." In the mid-1990s, the National Audubon Society had a membership of 570,000, down from a peak of 600,000 in 1990. Its operating budget was $46 million.[16]

More than any of the other big national groups, the Sierra Club has been a grass-roots operation in that its agenda has been driven in large part by its members and chapters. But according to executive director Carl Pope, during the 104th Congress, the club's leaders decided they had not been doing enough to get the membership actively involved in efforts to influence policy, as opposed to simply supporting what the leaders decided. "In donor-driven politics," Pope said, "people don't respond to public opinion; they respond to public energy." Recently, the club has been investing much more of its resources in community outreach, and the results have been "phenomenal." Sometimes, listening to the members has led the club's leaders in directions they did not want to go—a prime example was the club's 1998 referendum on whether to support restricted immigration on the ground that population pressures caused by immigrants were placing insupportable stress on the country's natural resources. The proposal was defeated, but the referendum issue used up a chunk of the club's $40 million budget and considerable staff time. Club membership has held steady at 500,000 for a number of years. It could be higher, but Pope said that it would cost more money to recruit more members and that the club's goal is keeping members loyal rather than expanding its membership rolls.

The big National Wildlife Federation (NWF), the "General Motors of the environmental movement," with an annual budget of some $100 million and membership of nearly 2 million—nearly 4 million if "supporters" are included—has also decentralized its operations. It is redeploying many of its resources to ten regional offices, and in a symbolic gesture, it moved its national headquarters from Washington, D.C., to nearby northern Virginia. "Building our grass-roots base is something we haven't done very well in recent years; we did it only on specific issues when it was needed," said NWF's president, Mark Van Putten. "You can't do it that way. You have to build it for the long run, and that involves skills training, identifying leaders, having regional activities and institutions." Van Putten also said that his organization and all environmental groups need to get back to the essential values embodied in Aldo Leopold's vision of the community of life: "We environmentalists have become too enamored of our own expertise, our ability to pronounce the names of all those chemicals. We need to say what we stand for and reclaim the moral high ground."

The Wilderness Society, one of the 20th century's storied conservation groups, had some "bumpy" years in the 1990s, going through three chief executives in four years and experiencing declines in membership and financing, according to its leader in the waning years of the decade, William Meadows. However, the organization righted itself, he said, and by the late 1990s it had a membership of 250,000—down from its high of 380,000—and an annual budget of $20 million. The society recently adopted a "strategic vision" centered on the creation of a national system of wildlands. Meadows emphasized that the society's strategy is based on a commitment to place—the idea that people care about and will want to protect specific places that they know or can visualize.

Place-based conservation has made land trusts and other organizations dedicated to preserving particular pieces of real estate among the most successful environmental enterprises in recent years. A number of states have adopted legislation and floated bonds to pay for acquisition of land to save dwindling open space. The Nature Conservancy, which acquires aesthetically appealing or biologically important lands to preserve them from development, thrived in the 1990s despite attacks on it by property rights radicals. "What we are seeing is strong dismay on the part of public about what is happening to open space," said Jean Hocker, president of the Land Trust Alliance, which supplies information and services to the rapidly growing number of regional and local land trusts around the country. The trend underscores the fact that people respond most readily to environmental problems where they live, work, and play.

Greenpeace USA, which had been regarded as the left flank of the mainstream American environmental movement and took on issues such as disarmament as well as saving whales, reversed the general trend toward jumping on the grass-roots bandwagon toward the end of the century. The organization's membership, which had reached 1.2 million after its ship the *Rainbow Warrior* was blown up by agents of the French government, plummeted to 500,000 in the mid-1990s. The organization's board of directors responded by drastically reducing the size of its staff and ending the organization's national door-to-door canvass to enlist new members and activate established members to support its campaigns. Greenpeace staff member Kalee Kreider, one of the bright "Young Turks" of the movement, said that the canvass had been losing money and not working particularly well in organizing people to support the campaigns. But former Greenpeace chief executive officer Barbara Dudley said that she thought the decision was "stupid and disastrous." The canvass may have been an inefficient way to raise money, she said, but it was the only way to find new members—aside from raiding other environmental groups. And, she added, "without grass roots, you are nothing." Without grass-roots support, she insisted, environmental groups have no clout in Congress or statehouses except for the dubious power of persuasion. Leaders of some other national environmental groups were dismayed by the Greenpeace decision. Brent Blackwelder, president of Friends of the Earth, expressed concern, saying that "the Greenpeace demise has left a vacuum in our effort to cultivate the grass roots." Blackwelder's own organization has focused heavily in recent years on tax-based and budgetary means of achieving environmental goals.

The View from the Grass Roots

Local activists themselves were not convinced that the heed paid by the national groups to the importance of grass roots in the late 20th century would be the answer to their prayers. "To tell the truth," said Lois Gibbs of the Center for Health, Environment and Justice, "we don't do a lot with the mainstream groups. We are about prevention and building for tomorrow—changing industry to achieve sustainable development, sustainable housing. They are about control—how much comes out of a stack, how much gets discharged into the water, how much gets buried in the ground. In order to win their goals, they have to compromise. We can't. We build our organizations from the bottom up. They build from the top down." The center itself has remained small in staff and budget, she said, because its mission is rebuilding democracy, not building itself. "We don't go parachuting into a community and then win or lose and go home. It is going to take people in communities saving themselves."

Terri Swearingen, who led the fight against the East Liverpool, Ohio, incinerator, also is unimpressed by the grass-roots commitment of the big national organizations. "They could have done more for us," she said. "I don't want to discount the achievements of the national groups, but they are not very effective. They operate more on theory than on passion. To achieve really meaningful change requires passion. Real energy for change is coming from the grass roots. When a person and their family are threatened, there is passion—they go out and fight like tigers."

Members of the grass-roots organizations are a potential army of shock troops for the environmental movement—and not just because of their numbers. In their ranks is an astonishing variety of expertise, ranging from land use management to gathering information about the effects of hazardous substances on human health to mobilizing local communities for political action. They are battle tested and deeply committed to their particular environmental causes. They are usually—although not invariably—uninterested in making concessions to wealthy donors or in acquiring perquisites for themselves.

To date, however, the grass-roots environmentalists have been more like isolated guerrilla units fighting small local skirmishes than an army trying to win a war. Despite efforts of the Center for Health, Environment and Justice to provide technical assistance and attempts by regional associations such as Great Lakes United to coordinate responses to environmental threats, the grass roots remain fragmented, their efforts uncoordinated. The national organizations still provide little assistance to the local groups and, perhaps more injurious to their cause, do little to solicit their knowledge and energy.

The Movement at Middle Age

An alleged loss of passion and militancy by the mainstream leaders and cadre is a recurring theme among critics from both within and without the environmental movement. Political scientist Michael Kraft said that "mainstream groups are losing ideological fervor to the greens and emerging grass roots environmental groups."[17] Deb Callahan, head of the League of Conservation Voters, which serves as the political arm of many of the mainstream groups, believes that the groups "are not ornery enough. We need warriors as well as peacemakers." Michael Frome, an environmental writer and longtime gadfly of the movement, asserted: "We need more than social reformers. We need revolutionaries. A lot of people in the movement are afraid of getting bloody noses. They are afraid of not getting invited to parties. That leads to compromising instead of fighting for what we know is right." Thomas C. Jorling, former head of the New York State Department of Environmental Conservation and former assistant administrator of the Environmental Protection Agency (EPA), contended: "The environmental movement has lost its vision. It has become a problem like the government." Tom Hayden, a California state senator and longtime voice of progressive causes, believes that environmentalism needs to reidentify with the spiritualism of nature in a return to what he calls in his book of the same name "the lost gospel of the earth."[18]

But the great majority of members of the big national organizations, both leaders and staff, are just as passionate and dedicated as they have ever been. What has changed is that as the organizations themselves have become more mature and professional, in many areas a certain bureaucratization and institutional fatigue have set in. In size and structure, many of the national and international groups have become like business organizations, with sizable budgets, plush offices, and reasonably attractive salaries and benefits. They are in a kind of comfort zone that makes them somewhat risk averse. The leaders of the national groups have become like the chief executive officers of big companies—educated, competent, and forceful—but, as former EPA administrator William K. Reilly, a former head of the World Wildlife Fund, commented, "they are not exciting." William Clark commented that every organization has two tasks. In this case, one is to protect the environment; the other, the institutional task, is to secure the life and well-being of the organization. The institutional task may in some cases be antithetical to the environmental task, as when, for example, a group avoids attacking corporate polluters for fear of offending a large donor.

Some critics also contend that those in the national environmental community are burdened by self-satisfaction, a sense that they are doing the Lord's work. David Hahn-Baker, an environmental consultant, college teacher, and former political director of Friends of the Earth, said: "My observation of myself and my colleagues is that we are environmentalists because it feels good. We are doing the right thing. We are right on the issues. When you radiate that, it can seem like arrogance and a projection of holier-than-thou superiority. It makes us less effective."

The legislative and legal victories won by the environmental groups in Washington and in statehouses in the 1970s and 1980s gained them a seat at the decision-making table and grudging respect from those who hold political and economic power in this country. To a certain extent, the environmentalists have become comfortable dealing with and accommodating to power. Some, in fact, eventually leave the movement to join government or business.

Acceptance by the entrenched power structure does not, however, make environmentalists into a special interest, as their enemies like to charge. Whatever their institutional character, environmentalists pursue goals—promoting clean air and water, protecting public health, preserving open space, safeguarding life support systems and the diversity of life—that are distinctly in the broadest public interest. Many of the lawyers, scientists, and other professionals who work in the national and regional environmental organizations could command substantially higher salaries in business or academia. As for local environmentalists, their only special interest is protecting themselves, their families, their property, and their communities from harm by those who use the environment heedlessly or for financial gain. It is a safe bet that most people who try to pin the "special interest" label on those in the environmental movement do so in their own special, usually venal, interest.

No Movement Is an Island

In January 1990, leaders of a group of organizations representing low-income African Americans and Hispanic Americans sent a letter to the mainstream environmental groups stating that "racism" and the "whiteness of the environmental movement" were the movement's Achilles' heel. The letter went on to say that the national groups needed to pay attention to the fact that polluting industries and other environmentally undesirable facilities such as toxic waste dumps and incinerators were habitually placed in "Third World communities" in the United States. It warned that "white organizations isolated from our Third World communities can never build a movement." The letter was a major milestone for what is now known as the environmental justice movement.

One of those who signed that letter was Richard Moore, coordinator of the Southwest Network for Environmental and Economic Justice, a binational U.S. and Mexican group that promotes cooperation and networking among grass-roots activists in communities fighting environmental racism. These groups try to protect themselves from such threats as pesticide contamination, lead paint in housing projects, uranium contamination, toxic waste dumps, and other pollution and to preserve their interests in the vital question of access to land in the Southwest. The network is also involved in creating and protecting jobs and in other economic and social issues such as economic development, improved working conditions for farmworkers, and low-income housing.

"The letter," said Moore, "was not about making accusations. We were not accusing anybody of racism. What we were trying to do was open a dialog. We invited national groups to our communities to taste the water, to see, to smell, to touch the things we are facing. We were making a big effort to build coalitions. We said, 'Let's do this together.'" Unfortunately, Moore said, "the national groups missed a golden opportunity. . . . They didn't get it together. They talked about too little time and too few resources. One of the reasons they didn't seize the moment is that they were stuck in the mode that they knew more than the rest of us."

Many environmental activists working at the local level complain that the professionals from Washington, New York, and San Francisco are willing to share their expertise and tell them what to do but seldom recognize the expertise of the locals or do what the grass roots tells them needs to be done. And despite honest efforts to redress the imbalance on the part of many of the nationals, their staffs and boards remain overwhelmingly middle-class (or richer) and white.

The great legislative and legal victories won by the national environmental organizations after midcentury provided essential tools needed by communities and local environmental organizations to fight their battles. By the end of the century, however, it seemed clear that the national organizations needed help from the citizens and organizations in local communities more than the communities needed them.

Strengths and Weaknesses

At the beginning of the new century, the environmentalists have many strengths to draw on as they prepare for the struggles just ahead. Chief among them is that they enjoy the support of a large majority of the American people, who want more, not less, environmental protection. Although their ethical fire has seemed somewhat dampened in recent years, environmentalists are entrenched on the moral high ground, and efforts to dislodge them, even by a popular president such as Ronald Reagan or by assaults from the political right, repeatedly fail. Politicians have discovered that it is risky to be overtly anti-environment. As John Flicker noted, "Environmental values are now in everybody's subconscious." The environmental organizations' leaders are battle hardened, and their staff rosters are filled with bright, well-trained, and dedicated professionals. They are now accepted by government and industry as a legitimate—if often irritating—voice in the conduct of the nation's affairs. An infrastructure of laws, treaties, and institutions to protect the environment is in place—sometimes loosely, it is true—from the global level to the local level. Despite their frequent disarray, the environmental groups have established workable networks for cooperation, not just in the United States but also globally.

Although publishers, editors, and television and radio network and station executives often slight the environmental story, either failing to cover it or giving it inadequate space or time, environmental news, reported by a growing number of talented and energetic reporters attracted to the stories, has become a part of journalism at all levels. Environmentalists themselves are becoming increasingly skillful at using the electronic media to put out their message and move citizens to action.

In academia, environmental history, environmental sociology, environmental policy, environmental theology, and, of course, environmental science and environmental studies have become flourishing disciplines. Although environmental education is in its infancy, more and more Americans are growing up with not only a knowledge of environmental issues but also a core belief in ecological values. A lot of kids who will be voters and decision makers early in the 21st century are growing up as environmentalists.

If the environmental movement has strengths to carry it into the next millennium, it also has weaknesses that must be addressed or its progress will be checked. The leadership of many of the organizations is aging. Young people coming into the movement tend to join the more radical organizations, not those in the mainstream. Greenpeace's Kalee Kreider warned that "we need to find some way to reach generation X." The Sierra Club's Adam Werbach contended, "It's time to evolve from an aging movement that reacts to emergencies to a youthful movement . . . powered by a broad vision for future generations."[19]

The American environmental movement seems to have no broad, shared vision of where it wants to take us. Kathryn Fuller, president of the World Wildlife Fund, stressed that the movement "needs a set of commonly agreed worldwide environmental objectives." The buoyant optimism of the 1960s and early 1970s, when young people stormed off campuses and into the environmental groups, intending to change the world, has evaporated. The professional environmentalists have become problem solvers, trying to solve one environmental problem or set of related problems at a time. In part, this probably reflects an overall reduction in confidence on the part of the American public in the future of the country and its economy. But it also suggests that today's environmentalists have not fully grasped the magnitude of their task, nor do they recognize their full power to create change. As David Korten observed, nongovernmental organizations "perceive their role as limited to the periphery."[20] But if the environment is to be saved, the environmentalists will have to play a central role in changing the world.

The environmentalists have not yet fully engaged the evolving tactics of their opponents in the corporate world, on the political right, and in the "wise use" movement. Corporations and, especially, their trade associations now deploy a broad range of weapons against the environmental agenda, from buying political support at all levels of government to creating bogus citizens' organizations to "greenwash" their activities to initiating lawsuits intended to intimidate local environmentalists, the so-called SLAPP suits (strategic lawsuits against public participation). Although a small but growing number of companies are coming to the conclusion that environmentally sound products and processes are in their long-term interest, many corporations spend large sums of money to counter the environmental movement, sums the environmentalists cannot match. Riley Dunlap found that a few think tanks and academics funded by the right-wing foundations and by industry "put out an enormous amount of disinformation" on issues such as global warming and are having a disproportionate influence in policy debates.

Many environmentalists have also been too quick to dismiss the significance of the "wise use" movement. Although it is true that many of the "wise use" groups are facades for extractive industries or vehicles for right-wing ideologues or opportunistic hustlers, the movement also represents real anger on the part of real people who feel personally victimized by the way environmentalism has affected their jobs and their communities. As Michael McCloskey noted, there is "a sense of alienation in small towns all around the country. People in lumber towns in the Pacific Northwest feel alienated and abandoned and powerless." It is no wonder, he said, that "for the first time, we have a grass-roots opposition."

The environmentalists are addressing an extraordinarily broad array of threats to the natural environment and human health. They have well-thought-out goals for what must be done to give us clean air and water, to protect land and species, to rid us of the insidious dangers of hazardous substances in the environment. They are beginning to set priorities for use of their resources on these problems. But they rarely come to grips with the root political, economic, social, and cultural causes of the problems.

The environmental movement is guided by John Muir's dictum that everything in the universe is hitched to everything else. It is firmly rooted in the land ethic of Aldo Leopold, which views humans as citizens of the entire community of life. Charges that environmentalists care only about wilderness and wildlife and not about human welfare are not only patently false; they fly in the face of the truth that the fate of humans and nature are inextricably bound together. When, sitting in his tiny cabin on Walden Pond, Henry David Thoreau wrote, "In wildness is the preservation of the world," his deepest fear was that encroaching industrial civilization would have disastrous consequences for both human welfare and the human spirit. Most environmentalists view their struggle to preserve what is left of wildness in the world as a service to humanity as well as to nature.

Despite this tradition, the mainstream environmental movement has yet to embed itself in the workaday human community—where people live, worry about their jobs, send their children to school, go to church and synagogue and mosque, and are exposed to myriad social as well as physical insults in their environment.

It is within this community, however, that much of the future of environmentalism will be determined.

Chapter 4
Environment, Community, and Society

Peggy Shepard was running for the Democratic Party district leadership in West Harlem in 1985 when several of her volunteer campaign workers asked whether she would be getting local people jobs in the municipal sewage treatment plant being built on the Hudson River at the western edge of the district.

"I didn't know anything about the plant," she recalled. "When I started looking into it, I thought the issue was jobs. But by the time the sewage plant was on-line, we discovered that the issue wasn't just jobs; it was an issue about an environmental nuisance. Then we looked around and found that we were also a dumping ground for bus depots, transfer stations, and other environmental hazards. People kept coming to me as district leader saying they had health symptoms and asking for help."

To fight back, Shepard and other citizens of the district formed an organization called West Harlem Environmental Action. With pro bono help from the Natural Resources Defense Fund and a big New York law firm, the organization sued the city of New York over the sewage plant and won more than $1 million in environmental benefits for Harlem.

But the new institution did not fold its tent after its victory. "We thought it important to not just respond to a crisis but to offer a vision of what a community can be," Shepard explained.

It soon became clear to the Harlem activists that underlying the issues of jobs and the environment was the broader issue of social justice—of economic and environmental inequity foisted on a community because its citizens were lower-income African Americans and Hispanic Americans. While the city was planning to build a major park on the affluent downtown Hudson waterfront, public housing and parking lots were being designed for Harlem's riverfront.

Today, West Harlem Environmental Action speaks out on environmental issues for all of northern Manhattan. It addresses a wide range of problems, including contamination by lead and other toxins, water quality, insect pests in houses, high asthma rates from air pollution, the need for open space, and many more. It is also involved in issues of community economic development, such as attracting needed retail stores, reclaiming brownfields, and creating jobs. "There is a real link between community development and community health that we try to make visible to all the people," Shepard explained. The organization publishes its own bimonthly newspaper and created a youth group called Earth Crew, which engages local young people aged thirteen to nineteen in environmental education, open space gardening, and other projects.

"The environmental agenda has to be broadened," Shepard asserted. "The mainstream groups have to understand there is an appropriate role in dealing with community concerns. They have to reach out for community partners."

In parts of the Pacific Northwest, as in far too many other places in the world, valuable things have been disappearing. Salmon have been disappearing. Trees and birds and topsoil have been disappearing. So have jobs and healthy communities. These valuable and necessary things have fallen victim to a swelling mega-economy, to runaway land development, to inappropriate technology and automation, to an expanding global market catering to an expanding global population. Environmentalists have tried to protect the salmon, trees, watersheds, and wildlife by lobbying for stronger laws and by trying to enforce the laws through the courts. But they have paid relatively little attention to jobs, to communities, or to the other building blocks of a decent, livable society.

In 1991, Spencer Beebe, former vice president of The Nature Conservancy, formed an organization called Ecotrust to try a new kind of environmentalism. Its goal, Beebe said, is "to integrate conservation and development by building on the cultural and economic traditions of local communities." It addresses the environmental, social, and economic well-being of a local community as inextricably interwoven.[1] One of its projects was the creation of the Willapa Alliance, an organization of farmers, fishermen, a logging company, tourism businesses, and others who live near the Willapa Estuary of southern Washington State and earn their living from the estuary's ecological resources. The estuary is still one of the cleanest in the country, but it is being steadily degraded by removal of old-growth forest, draining of wetlands, use of biocides, and discharge of sewage into the estuary, along with a number of other ecological insults. The area was also one of the poorest in the state because, Beebe said, it had failed to make the transition from an industrial economy based on foundries, wood mills, and large-scale logging and fishing. With the help of capital provided in part by the Shorebank Corporation, a Chicago banking institution with a long history of assisting community development, the alliance is seeking to build a new economy of permanence based on conservation and careful use of its salmon and oysters, its trees, its soil, its streams, and the rest of its natural capital.

"The old environmental movement is over, in a sense," Beebe said. "That movement arose as a defense against the industrial economy and to save some precious pieces of the landscape from human industrial endeavor. It was appropriate. But we need now to move to a new era where we find synergy and sympathy between the built and natural environments. We need to move from a strategy of defending bits and pieces of nature to recognizing the links between a healthy community and a healthy environment."

This sentiment was echoed by Rose Augustine, president of Tucsonians for a Clean Environment. After years of fighting to force the cleanup of toxic contamination from a Hughes Aircraft Company plant that was seeping into the drinking water of her largely Latino community on the south side of Tucson, Arizona, Augustine came to see that the environmental problem was far more complex than she had realized. "We should be looking to expand the definition of environment," she said. "Children are the most precious resource we have, and things like crime, gangs and drugs are part of the environment and people had better stop and look at it that way."[2]

What is the environment? It is the air, the land, the water, the climate, the weather. It is grass and trees and animals. But it is also a *place,* a place in this human-dominated world where people live, work, play, build houses and stores and factories, pray, send their children to school. It is a place where they grow crops, raise livestock, climb mountains, sit on the beach. It is where they spend their money, read newspapers, and watch television, where they vote or do not vote, belong or do not belong to civic institutions. It is where they feel secure or endangered in their homes and on their streets. The environment is a created as well as a natural community.

As Aldo Leopold and others have instructed us, humans are part of the entire community of life. But we also belong to particular communities of place. Since we first saw the photographs of a small, blue planet Earth from space, we have begun to think more seriously about a global community. In many important respects, there are national and regional communities linked by common geography, history, economy, and ideals and systems. In its most workable form, the community is a local place—a village, a town, a city, a suburb, an urban or rural neighborhood.

The kind of environment we live in—and will leave to our children—depends on the kind of *society* we create within our communities. The social infrastructure fundamental to a healthy environment in this human-dominated world includes not only good laws and public institutions; a thriving, rational economy; and a responsive political system but also shared information, knowledge, goals, and values; active civic organizations; and, crucially, mutual tolerance and regard among the citizens of a community and concern for one another's well-being.

William Shutkin, founder of Alternatives for Community & Environment, a Boston-based civic group that works chiefly with low-income communities, contended that environmental problems in the United States are in large measure a result of a widespread deficit in *social* infrastructure. Dana Lee Jackson, cofounder of The Land Institute in Salina, Kansas, said she believes that the ecological challenge of this era "is to transform our society so we can act on our ecological knowledge, change destructive patterns and develop a sustainable society."[3] In *Habits of the Heart,* a penetrating analysis of American society at the end of the 20th century, Robert Bellah and his coauthors draw a parallel between the destruction that modernity and rampant individualism in the United States have wreaked on the natural ecology and the destruction they have inflicted on our "social ecology." Unless we begin to repair our social ecology, they warn, "we will destroy ourselves."[4]

If we can save the environment only by repairing our society and our communities, the environmentalists will fail because, at least until now, they have been focusing almost exclusively on the physical environment. Even at the local level, most grass-roots environmental organizations, unlike West Harlem Environmental Action, concentrate their efforts on the immediate problem—the waste dump, the petrochemical plant, the incinerator, plans to log the stand of ancient redwoods. The underlying flaws in our social systems that cause or contribute to the environmental predicament are rarely addressed by environmental organizations or the environmental movement as a whole.

Michael McCloskey, one of the more experienced and thoughtful of the mainstream environmental leaders, expressed the view in a paper prepared for the Sierra Club's board of directors that the club could take positions on social issues but only within certain "boundaries." Those boundaries are restricted by, among other things, the institution's legal charter, a membership who signed on for specific goals, fiduciary responsibilities, the danger of diluting the club's effectiveness, and the danger of the club's being seen as a "predictable and uninteresting general-purpose progressive organization with a green specialty," McCloskey wrote. The club, he added, should insist on a "logical nexus" between social issues and environmental ends. He sees such a nexus with social equity issues because maldistribution of wealth strains the resource base. He also sees a link between the destruction of jobs and the environment by automation technology.[5]

Other environmentalists are less open to addressing a range of social issues. Some leaders say that although they are personally committed to social causes such as more equitable distribution of wealth, their organizations do not have the money or staff to deal with issues outside their own agendas. "We are not equipped to solve the social problems of the country," said Gaylord Nelson, counselor for The Wilderness Society, former senator from Wisconsin, and founder of Earth Day. Others will admit, usually off the record, that the wealthy donors on their boards would not accept progressive initiatives or that they could not afford to offend conservatives among their dues-paying members.

But some who give thought to these issues have concluded that without addressing the underlying causes of the environmental dilemma, environmentalists are merely attempting to treat its symptoms, an attempt that inevitably will be overwhelmed by the onrushing tide of people, poverty, special interest politics, market economics, and technology. Peter Raven of the Missouri Botanical Garden stated: "The environmental movement should be approaching sustainability as its main goal. If you accept that the movement has responsibility for social justice and systems that will be sustainable everywhere, including urban areas, then it is now nibbling at the edges of the problem."

A few of the newer groups, such as West Harlem Environmental Action, Ecotrust, and Alternatives for Community & Environment, are seeking to integrate environmental and social agendas. Several mainstream groups, including the Natural Resources Defense Council (NRDC) and the Sierra Club, have made forays into the arena of social reform. But by and large, as William Shutkin noted, "you are hard pressed these days to hear our national environmental groups talk about democracy, or talk about social and civic infrastructure, even though these are the critical linchpins of successful environmental outcomes."

If, however, environmentalism is to fulfill its mission of preserving and enhancing our habitat, it must reach its potential as a mass social movement and acquire the means and energy to do so. That cannot be done by nibbling at the edges of pollution and resource depletion. National organizations and local activists alike will have to dig to the roots of the problem, which lie in our economy; our politics; our science; our race, class, and ethnic relationships; our schools; and our churches, synagogues, and other social and civic institutions.

Social ecology broadly describes the interrelationship of natural and human communities. But it is defined in different ways from different perspectives. The concept of social ecology presented by economist E. F. Schumacher in his book *Small Is Beautiful* and developed by Murray Bookchin envisions a departure from the giantism of our current economic and political systems to a system of decentralized local economies that rely on small-scale technologies in communities ruled by direct participatory democracy. Brian Tokar, a social ecologist and teacher, describes social ecology as "the most coherent expression of ecological radicalism, and an approach uniquely suited to confronting the challenges of the next century."[6] Social ecology is not radical, however, in the context employed in *Habits of the Heart,* which has to do more with morality and the relationship between individuals and society. The definition of social ecology used by Bellah and coauthors is thus somewhat different from that used by Tokar and Bookchin, who do see social ecology as radically reordered economic and political arrangements. *Habits* envisions a return to a more traditional approach to community, a cooperative tradition in which the common good is not overwhelmed by selfish individualism—hardly radical.

Obviously, environmentalists cannot address every social issue or seek to mend every flaw in our institutions, habits, and customs. But there are pressure points that are an obvious nexus between the social and environmental missions. There are also existing social and political movements with which environmentalists can form alliances and partnerships, either permanently or on an ad hoc basis, to fight for basic reforms. National environmental groups can bring their skills, professionalism, contacts, and clout to bear on the pressure points, particularly if they act as agents and conduits for community-based environmental activists. Grass-roots organizations can move from fighting a specific environmental threat or insult to organizing themselves to change the conditions in their community that permit such threats and insults.

Community-Based Development

There is another movement out there. It is called community-based development, and its mission is to make the country's communities better places by building their economies, creating jobs, creating and restoring housing, dealing with crime and drug issues, attending to community health, and generally seeking to improve communities' physical and social infrastructure. It, too, is a mass social movement, although it attracts far less public attention than does the environmental movement. At its core are an estimated 5,000 to 7,500 community-based development organizations, the majority of them located in low-income inner-city and rural neighborhoods, most often with minority populations, and usually created and led by residents of those neighborhoods. But the community development groups also have enlisted the assistance of government at all levels; nonprofit intermediaries that provide technical assistance and funding; academia; and private sector institutions, including banks and corporations. Some 300,000 people are believed to be employed in the various aspects of community development.[7]

Just as the environmental movement has typically ignored (and occasionally opposes) community economic development, the community-based development organizations have paid little heed to environmental issues. However, that has started to change in recent years. One reason is that the burgeoning environmental justice movement, discussed later in this chapter, has demonstrated that the poor, minorities, and residents of inner cities bear disproportionately heavy burdens from pollution and other environmental insults. Probably more important is that development professionals have come to see that a degraded environment is a serious obstacle to building a local economy and that, conversely, cleaning up the environment not only produces a positive climate for economic growth but also can create jobs for community residents.

Alice Shabecoff, an expert in the field of community development, wrote: "Environmental protection and economic development, frequently described as contradictory goals, have in fact proved mutually reinforcing at the neighborhood level. . . . The environmental initiatives developed by community groups are prime examples of 'sustainable development,' that is, development that clearly takes into account three elements—the economy, ecology and community."[8]

Among the market-oriented environmental projects initiated by community development organizations are brownfield rehabilitation, energy audits and corrective repairs, waste cleanup, remanufacturing of recycled materials, real estate improvements, eco-industrial parks (manufacturing sites designed and operated to preserve the environment), and other ecologically sensitive commercial development.[9] Communities increasingly are realizing that, as David Hahn-Baker, a consultant and longtime environmentalist and social justice advocate, stated, "health, community development and environment are integrally linked to one another; community development is not just about numbers of new housing units or jobs created but about, among other things, human health and environmental health."[10]

In recent years, the national environmental groups have issued a stream of mea culpas over their failure to "reach out" to grass-roots organizations and have pledged to do better in the future. But their "reaching out" is often little more than an effort to enlist the political support of the grass roots for their own programs. As Richard Moore of the Southwest Network for Environmental and Economic Justice pointed out, unless the larger groups' efforts are planned from the start in collaboration with local communities, they may work against the communities' economic interests. As an example, he cited conservationists' efforts to limit logging in the Southwest that would affect the firewood traditionally used by the area's poor Hispanic American and Native American residents for cooking and heating.

Michael Fischer, a longtime environmental leader and former executive director of the Sierra Club, is convinced that "the success of the environmental movement depends on merging or meshing with the community development movement. The way to do it is to go to those folks and say, 'What are your needs? What is your situation? How can we help?' But now 'outreach' means giving the truth to you poor, ignorant folks."

In turn-of-the-century America, both movements need each other. The community development professionals, having come late to their understanding of the links between the environment and sustainable economies, need the knowledge, experience, and institutional capacity of the environmentalists to help them deal effectively with these issues, not to mention the clout the environmental movement exercises with the public and politicians at the national and state levels. As the President's Council on Sustainable Development suggested, environmentalists can work with local communities on long-range development planning that incorporates ecological needs and can work with communities in the execution of such plans.[11] The environmentalists have access to pools of capital that can build the kind of enterprises needed for real development—for example, as discussed in the next chapter, NRDC was able to attract the many millions of dollars needed to capitalize a paper-recycling mill in the South Bronx that it sponsors along with Banana Kelly, a community development group. Environmental organizations can help the community-based organizations build institutional capacity.

And in an era of devolution, deregulation, and interest group politics, the environmentalists will need allies and partners in local communities both to help devise and to help carry out an environmental agenda. The community development organizations could help the environmentalists with locally oriented market strategies. They could help the environmentalists to expand their horizons and agendas, to "think more imaginatively about their missions and methods." As David Korten noted, such local groups "push social self-assessment, experimentation and change, reflecting evolving values of people. They are social and political catalysts."[12] Moreover, the environmental movement has yet to make an adequate response to opponents who charge that they do not care about the fate of jobs and communities as long as they achieve their own goals. Even though these charges are exaggerated and sometimes scurrilous, there is enough truth in them to explain the alienation of many Americans—workers, loggers, ranchers, miners, fishermen—who ought to be the environmentalists' allies in a new "blue–green" coalition but instead have become recruits of the so-called wise use movement.

Unfortunately, the environmental and community development movements, seemingly natural allies, barely communicate with each other. Indeed, sometimes they seem like two big ships sailing a few miles apart but enclosed in dense fog and unaware of each other's existence. There are a number of explanations for this mutual lack of comprehension and communication. Many in the community development field perceive environmentalists as opposed to economic development. It is a perception reinforced by the fact that it is occasionally true: environmental groups do sometimes oppose economic development projects on the ground that they are ecologically destructive. There is no doubt that class antagonism between the middle-class, white national environmental cadre and the predominantly working-class, racially and ethnically diverse community development movement exists. The environmentalists often do not understand the dynamics of community relationships—NRDC, for example, has had endless difficulty in reconciling the demands of competing community groups in the South Bronx. For their part, community groups often view environmental projects as having no value unless they produce buildings or jobs.

Only recently has a dialog begun to open among imaginative thinkers within the two movements. John Berdes of Shorebank Enterprise Pacific, a corporation established by Ecotrust and the Shorebank Corporation to fund the sustainable use of raw materials in the Pacific Northwest, noted that the movements "come out of wholly different cultures and languages" and added, "What we're creating here is a new culture and language emerging around what we call the conservation economy, which embraces ecology and economics as much as it embraces economy and equity."

Perhaps in the coming years, these two ships will sail in the same direction and toward the same goals.

Environmental Justice

"The environmental justice movement," explained Deeohn Ferris, one of its leaders, "is the confluence of three of America's greatest challenges: the struggle against racism and poverty, the effort to preserve and improve the environment, and the compelling need to shift social institutions from class division and environmental depletion to social unity and global sustainability."[13]

At least that is what environmental justice ought to be about. The movement began in the 1980s when civil rights leaders and institutions in the United States began looking around and noticing that an unfairly heavy share of the country's environmental insults were being heaped on the poor, most often poor African Americans, Hispanic Americans, and Native Americans. A 1984 report of the Urban Environment Conference found that "minorities are the targets of a disproportionate threat from toxins, both in the workplace, where they are assigned the dirtiest and most hazardous jobs, and in their homes, which tend to be situated in the most polluted communities." A 1987 study by the United Church of Christ's Commission for Racial Justice found it "shocking to discover that African Americans, Hispanic Americans, Asian Americans, and native Americans disproportionately live in communities with a dangerous concentration of hazardous waste sites. . . . Even more outrageous, we have found that this reality is no accident, no mere random occurrence. . . . It is, in effect, environmental racism."[14] As discussed in the previous chapter, a 1990 letter from civil rights and community organizations pointedly called the attention of the mainstream environmental groups to the whiteness of the green movement. A year later, the First National People of Color Environmental Leadership Summit was held in Washington, D.C. The organizers described the summit as a "pivotal step in the crucial process whereby people of color are organizing themselves and their communities for self-determination and self-empowerment around the central issue of environmental justice." The goal of the conference was to "reshape and redefine the American environmental movement."[15] As Robert D. Bullard, an intellectual leader of the environmental justice movement, explained, the movement is aimed at putting "participatory democracy" to work for communities and for the environment.[16]

Responding to the criticism, the national environmental groups were quick to concede their failure to pursue environmental equity and to accept the precepts of environmental justice. They pledged to make a better effort to enlist people of color in their staffs, their boards, and their membership rolls, and they seemingly made an honest, if not very successful, effort to do so. Minorities are still very much minorities in the mainstream groups.

Environmental justice advocates can do without the traditional green groups, although they could do better with them. The point here is that the abuse of our environment is of a piece with and springs from the same flaws in our society and culture that cause racism, economic inequity, unequal distribution of political power, and many of the other forms of injustice that plague our democracy at the turn of the century. The poor often must live in congested city neighborhoods with heavy concentrations of air pollution, in decaying housing stock contaminated by lead and other toxins. Or they live in rural backwaters with neglected infrastructure, often with degraded water supplies and inadequate diets. The schools of the poor tend to be substandard. Because of low incomes and poor education, the poor are often politically apathetic. Moreover, they cannot make political contributions. Politicians can dump environmental and social insults such as waste treatment plants and toxic waste dumps into their communities without fear of reprisal. And, of course, people of color make up a high proportion of the poor in this country. Unless the mainstream of the environmental movement not only embraces the principles of environmental justice but also directs its energies to ridding society of the flaws that gave rise to it, they will be unable to deal effectively with the looming dangers to the environment.

Joining forces with the environmental justice movement would link the environmentalists with the quest for civil rights and social justice in the United States, thereby helping to revivify the environmental movement's flagging moral authority. The quest for environmental justice is already fueling the civil rights movement with new energy. Congressman John Lewis, a pioneer of the civil rights movement, asserted not long ago: "Clearly, the goals of advocates of social and economic justice and the goals of environmentalists have begun to converge. This shared vision makes for a stronger movement when diverse groups, organizations and communities view environmental protection as a *right* of all, not a *privilege* for a few" (emphasis in original).[17]

An alliance with environmental justice groups would also engage environmentalists far more than they are today in the reclamation and rehabilitation of our cities, particularly our inner cities, which are often degraded by some of the worst physical environments in the country. Although the efforts of environmental organizations to fight pollution and change patterns of energy use and transportation do help the inner cities, such results are usually an unintended by-product of the environmental mission. Environmental programs specifically designed to mesh with urban rescue programs, such as NRDC's project with Banana Kelly in the South Bronx, are rare. The environmental movement is, with some reason, regarded by poorer city dwellers as run by and for suburbanites. Community organizer Ernesto Cortes Jr. noted: "People in modern industrial societies, particularly those living in the cities, are atomized and disconnected from each other. Particularly in the suburbs, far too much of the American search for 'fulfillment' is centered on the individual, making his or her relationships utilitarian and narcissistic in nature. This fragmentation leaves people increasingly less capable of forming common purpose and carrying it out."[18]

A close partnership between the environmental and environmental justice movements would do much to pull these social fragments together. As Morris (Bud) Ward, head of the National Safety Council's Environmental Health Center, pointed out, "The inner-city poor remain estranged from the environmental movement because the movement has not reached out to them." But the environmentalists do not seem to have recognized even the utilitarian reasons for an alliance: that by helping reinvigorate the communal life of our cities, they would be helping relieve the commercial, industrial, residential, and recreational pressures that are rapidly eating up the American countryside.

Finally, it is well to keep in mind that environmental injustice is relative. Although the poor, people of color, and the politically neutered have the worst environmental insults heaped on their communities and their places of work, every one of us—every American, every person in the world—is victimized by those insults to some degree. As the bumper sticker says, "We all live downwind." All of us are in need of environmental justice.

Labor and the Environmental Movement

The relationship between the environmental movement and organized labor is complex, wary, sometimes hostile, and usually unsatisfactory. In theory, unions and environmental organizations ought to be allies. Workers suffer more than their share of environmental abuse; workplaces often house some of the most dangerous and ubiquitous environmental hazards. The Occupational Safety and Health Administration, which along with the Environmental Protection Agency and the Department of the Interior is one of the major environmental arms of the federal government, is part of the Department of Labor. Both environmentalists and union leaders often find themselves in adversarial relationships with corporations; corporate managers use many of the same tactics they have long used to weaken union power to weaken environmental regulation and the power of environmental institutions.

Both movements have relied heavily on federal legislation and enforcement to achieve their goals. On specific issues, such as protecting workers from the effects of pesticides and other chemicals, they can be on the same side and work together.

In practice, however, the two movements are often at sharp odds. Although union leaders and the rank and file may support environmental goals, their overwhelming interest is in job security, economic growth, wages that can support their families, and stable communities. Anything that threatens those interests or even has the potential to do so is a red flag waved in front of labor organizations—and with good reason. Few environmental threats can outweigh the immediate and urgent need of workers for jobs and a decent wage, and union leaders cannot take actions perceived by their membership as endangering their livelihood. Environmental organizations, meanwhile, have long focused single-mindedly on their anti-pollution, resource-protection goals without undue concern about jobs or communities. That has started to change in recent years, but the environmentalists continue to press their agendas with the argument that a healthy environment is necessary for a healthy economy. Although that is certainly true over the long run, the long run does not help a jobless worker with a family to feed and rent or a mortgage to pay.

A case in point from the late 1990s is the Kyoto Protocol to the United Nations Framework Convention on Climate Change. The treaty is a top priority of the environmental movement, which has pulled out all the stops in gaining support for it. To the argument that the treaty would have disastrous effects on the economy, put forward largely by the fossil fuel industry and conservative think tanks and politicians, the environmentalists have rightly replied that the economy and the nation will suffer far more over the long run if nothing is done. But the economy inevitably would feel some effects from the transition away from coal and oil. More important, some groups of workers, such as coal miners, could be devastated by the shift. That is why the American Federation of Labor and Congress of Industrial Organizations (AFL-CIO) is opposing the protocol. Jane Perkins, environmental liaison for the AFL-CIO, noted: "The mine workers are very vocal in their opposition to the protocol. The climate change issue challenges the very existence of the union."

Perkins, who came out of the union movement to do a stint as president of Friends of the Earth and then joined the labor federation, said: "A lot of the anger directed toward environmentalists is actually misguided—but it is real. The workers who have suffered the greatest loss of jobs and security are the ones who have the strongest feelings, like the timber workers and organized miners. You cannot have a discussion on how automation and technology are more of a threat to their jobs. They don't recognize that. It is the spotted owl and clean air legislation. It is visceral."

An increasing number of unions have been trying to find answers to the environment–jobs issue in recent years. At Perkins's request, John Sweeney, president of the AFL-CIO, appointed an advisory board from member unions to formulate an approach. Perkins said that the board asked her to come up with a set of principles "that say, if you are a trade unionist, what guides you when you are faced with environmental questions and issues." Four principles were adopted. The first principle is protection of the economic rights of workers and their families, "which, after all, is the reason for unions," Perkins said. The second principle calls for finding the highest common denominator for environmental protection consistent with the economic well-being of workers. The third calls for ensuring a "just transition" if environmental needs require changing the conditions of work; the fourth, using "common sense" and arriving at solutions by means of a democratic process. Environmentalists should have little difficulty adhering to such principles in their dealings with unions and workers.

The advisory board has also undertaken an effort to come up with a list of organic chemicals that persist in and pollute the environment, the existence of which no trade union could defend. The point of the exercise is to work out a way in which a labor union could "advocate banning stuff whose members make a living making that stuff," Perkins said. "This is revolutionary. How do you deal with the economic fallout, with the worker fear that goes with doing the right thing, and not make this stuff? That is so brave for the labor movement. We may not be able to do it, but we are trying."

The environmental movement needs to be braver in seeking common ground with unions and their members. Lois Gibbs, whose Center for Health, Environment and Justice has been working with local citizens' groups and unions to get papermaking companies to stop using chlorine in their manufacturing processes, said that it takes patience, imagination, and understanding of workers' needs and fears to win them over. "It has taken us ten years to build the trust of rank-and-file workers," she said. "We had to show them we were not trying to shut down papermaking plants. We had to work with them to find solutions." In one case, she recounted, her organization supported the floating of an industrial revenue bond to help the Boise Cascade Corporation finance the new equipment and other production facilities needed to shift away from a chlorine-based process.

"The mainstream environmental groups," Gibbs asserted, "don't understand that because workers care about their jobs, their families, their piece of the American dream, they also care about the environment. . . . It's all about building relationships."

Education and the Environment

Looking back in October 1998 at ten years of funding environmental projects, Edward Skloot, executive director of the Surdna Foundation, concluded: "Most American corporations and individuals, as well as their government representatives, still believe that natural resources and habitat are there for the taking. Without reformation of this dominating mind-set, and agreement on a few universal truths that give our individual and collective lives meaning and sustenance, we will not be able—and should not be able—to live sustainably in the world."[19]

How can "this dominating mind-set" be reformed? The question virtually answers itself. The shaping of minds begins first with immediate family, of course, but most profoundly with education, with the schools.

The religious right knows this well—that is why the conservative fundamentalist churches have sought to gain control of school boards and to shape their curricula and reading lists. In fact, gaining control of the educational system is a chief goal of the anti-government extremists of the right. As education consultant Ann Bastian wrote, "Privatizing public education is the center piece, the grand prize, of the right wing's overall agenda to dismantle social entitlements and government responsibility for social needs."[20] At the same time, corporations that exploit public resources or despoil natural systems are also engaged in what author Bruce Selcraig described as a "far-flung campaign to dumb down environmental education." Selcraig reported, for example, that the Exxon Corporation spent $1.6 million to develop and distribute a curriculum that somehow fails to mention any connection between petroleum and pollution. The American Coal Foundation's *Power From Coal* workbook instructs students that "more time" is needed to determine whether there is any connection between burning of coal and global warming.[21]

There is a lesson here that the environmentalists could well heed. It is in classrooms, from preschool and kindergarten through graduate school, that a major part of the struggle to achieve a sustainable, livable habitat and society in the 21st century will be won or lost.

Many public school systems now include environmental courses or at least environmental subjects in their curricula, and their number is steadily growing. Young children come home from school and teach their parents the basic environmental facts at the dinner table. Colleges now offer courses in ecology and environmental science, and their graduate schools turn out young men and women trained in environmental law, environmental engineering, environmental sociology, ad infinitum. Even business schools are increasingly offering courses in "green management."

By and large, however, environmental education remains shallow and manifestly inadequate for the challenge of preserving our environment in the 21st century. Daniel C. Esty of Yale University commented: "The extent to which we as a society have not made the right environmental choices shows that we don't understand our environmental resources. That is why environmental education is so important." As David Orr, chair of the environmental studies program at Oberlin College, observed, the educational system in the United States is devoted to training people to compete in the global economy but not to addressing the looming threats to the environment. Orr laments that "we continue to educate the young for the most part as if there were no planetary emergency."[22]

Anthony Cortese, president of Second Nature, a Boston-based organization dedicated to inculcating the importance of "education for sustainability" in professionals in higher education and improving environmental education in colleges and universities, contended: "The problem we have is not *in* education but *of* education. It is the people coming out of our most prestigious business schools and law schools and engineering schools who are leading us down an unsustainable path." The current educational system, Cortese said, reinforces "the deep cultural belief that humans are separate from nature. Secondly, instead of being taught in holistic and interdisciplinary ways, we are taught in separate little compartments—math is separate from physics, is separate from reading, is separate from chemistry. . . . The problem is we don't make connections between economics, biology, politics, engineering. This failure to engage in interdisciplinary and systemic thinking is what gets us into trouble."

Cortese, former dean of environmental studies at Tufts University, is trying to persuade school administrators that instead of having an environmental course here and there, they should integrate ecological and environmental material into every element of their curricula.

David Orr, whose students are working with him in designing and constructing a fully ecologically sound school building on the Oberlin campus, a building that several critics have called one of the most innovative architectural and design projects of recent years, asserted that the educational system needs to be restructured to produce ecologically literate citizens. "A genuinely liberal education will produce whole persons with intellectual breadth, able to think at right angles to their major field; practical persons able to act competently; and persons of deep commitment, willing to roll up their sleeves and join the struggle to build a humane and sustainable world."[23]

If, as Lester Milbrath tells us, we are going to have to "learn our way out" of the ecological and evolutionary cul de sac we wandered down in the 20th century, the environmental movement will need to have a far greater influence on our schools and our educational system than it does now.[24] It will need a greater voice on school boards, state regents systems, university boards of trustees, and other institutions that set educational standards as well as a greater voice in determining curricula and syllabi.

Environment and the Media

Even an educated citizenry cannot make informed environmental choices if it lacks the information to do so. As Konrad von Moltke of

Dartmouth College observed, "It is the role of the environmentalists to articulate the issues clearly." But they also must be able to deliver the message, and the principal way of doing so to an audience beyond their own members is through the mass media—television, radio, newspapers, magazines, and, recently, the Internet.

With media-savvy newer organizations such as the National Environmental Trust and Environmental Media Services, the ability of the environmental movement to present its message to the public is steadily increasing. There is still, however, much room for improvement. The big national groups often put too few resources into media outreach, and their messages are often designed to gain institutional publicity rather than to alert the public to a pressing issue. Grass-roots environmental campaigns are often hampered by a lack of expertise in and resources for getting their stories into the media—and they usually get little help from the nationals.

The chief obstacle to getting the environmental story to the public, however, lies not with the environmentalists but with the media. Most news organizations habitually ill serve their communities, whether the national community, big cities, or small towns, by failing to report environmental news on a regular basis and, even worse, by failing to understand how environmental issues should be integrated into all aspects of their coverage, from business news to national security. "The media are part of the problem," said Arlie Schardt, executive director of Environmental Media Services and former correspondent for *Newsweek* and chief executive officer of the Environmental Defense Fund. He noted that although reporters are interested in covering the environment, editors are all too often reluctant to schedule such stories.

The number of reporters covering environmental stories has increased dramatically since the 1970s. In the mid-1970s, only a handful of journalists wrote regularly about these issues. Today, the Society of Environmental Journalists, which was formed in the early 1990s to promote and upgrade environmental reporting, has well more than 1,000 members. But the amount of environmental news carried by the media has not been growing in recent years, and that carried by the television networks and many newspapers has declined substantially. The news organizations tend to focus on environmental news when there is a crisis, such as the 1989 *Exxon Valdez* oil spill or the disastrous chemical spill at the Union Carbide plant in Bhopal, India, in 1984, while paying little heed to stories such as urban sprawl, pollution of central cities, decline of biological diversity, and many other issues of vital importance to local communities, to the nation, and to the viability of the global environment. Even worse, environmental reporters are seen as suspect at many news organizations by editors or publishers who assume that those who cover the environment are environmentalists. This attitude, of course, has a chilling effect on hard-hitting environmental reporting.

A number of reasons have been advanced for the poor performance of the media on environmental issues. One is pressure from advertisers who are hostile to environmental regulation and to environmentalists. Another is that the senior managers of news organizations are no different from managers of other corporations who object to laws or institutions that interfere with their management decisions. Some media managers have succumbed to the myth that protecting the environment leads to economic decline and thus would hurt their company's revenues. There is undoubtedly some suspicion by media managers that environmental stories do not sell. That belief is not supported by available data, however. For example, a survey taken by the Roper Center for Public Opinion Research for the Freedom Forum's Newseum in 1996 found that 59 percent of respondents were "extremely" or "very" interested in environmental news, ranking the environment just below local news and crime but well above sports, national and international affairs, business and money issues, and politics.[25] It is more likely, however, that the failure of the media springs in large part from ignorance of the issues and poor judgment by media managers. Although some of them have become enlightened, many do not have a clue about environmental issues, a nontraditional journalistic subject, and show little inclination or intellectual resilience to learn about them. They seem unable to grasp the simple truth that the environment is the field on which all other issues are played out, from business and industry to national security to civil rights and even sports and culture. This may change when a new generation takes over the media, but it is a serious problem now.

Another important task for environmental organizations and educators, therefore, is to find ways to convince the lords of the press that they are ill serving their audiences and themselves unless they give the environmental story the full attention it merits and give their environmental reporters the same trust and latitude they give their business reporters.

Environment and Religion

Churches and synagogues have always been important centers of communal life in the United States, as elsewhere. They are where members of the community can look for moral and ethical guidance. But the churches and religion historically have not been active in addressing the moral and ethical dilemmas of the environmental crisis, nor have they sought to lay a spiritual foundation for cherishing and preserving nature. On the contrary, they have been regarded by many scholars as part of the problem.

More than thirty years ago, in a landmark essay in *Science,* historian Lynn White Jr. wrote that the Judeo-Christian tradition "bears a huge burden of guilt" for the degradation of nature by Western science and technology. That tradition placed humans, as the agents of an anthropocentric God, apart from and above nature and gave them dominion over and authority to exploit the natural world.[26] More recently, theologians Mary Evelyn Tucker and John Grim noted: "For the most part, the worldviews associated with the Western Abrahamic traditions of Judaism, Christianity and Islam have created a dominantly human-focused morality. Because these worldviews are largely anthropocentric, nature is viewed as being of secondary importance." They also pointed out that these religions have been concerned more with personal salvation than with responsibility for preserving the natural world, which is often viewed as corrupting.[27]

Of course, the Judeo-Christian Bible also admonishes people to be good and faithful stewards of the natural world created by God. But this tradition has been slighted by the churches and their clergy for centuries. Eco-theologian Thomas Berry believes that a "split occurred between the secular and spiritual worlds" in the 14th century when the Black Death wiped out a large proportion of Europe's population and the survivors concluded that God was punishing the world and they had best seek redemption outside of the natural world. This disengagement of secular and spiritual concerns, wrote Berry, founder of the Riverdale Center for Religious Research, "allowed industry and commerce, with the assistance of science and technology . . . , to seize control of the natural world and to exploit it."[28]

Until recently, however, there was little sign that religious institutions were giving what Tom Hayden, former student radical, current California state senator, and all-purpose progressive, described as "the kind of passionate engagement in the environmental cause that the clergy of America gave to civil rights in the 1960s, or that the priests in Latin America invested in Liberation theology on behalf of the poor." What we are seeing today instead, he wrote, "is the Religious Right condemning environmentalists as pagans while defending the property rights of polluters as somehow protected by the mandates of Genesis."[29] It is mystifying that the Christian right opposes efforts to protect the environment and thereby accedes to or even collaborates in the destruction of the works of the Creator.

In the latter part of the 20th century, faced with unmistakable evidence of a dangerously eroding natural environment, religious institutions began reexamining their attitudes toward and teachings about the relationship of humanity, nature, and God. Theologians of many creeds began seeking a new cosmology that resacralizes nature and restores humans to their place as part of a natural universe in which God resides.

As the millennium drew to an end the world's religions clearly were beginning to address the environmental crisis with, well, religious zeal. A front-page article in the *Los Angeles Times* observed: "Some activists call it the birth of a religious movement as significant as the battle against slavery. Churches, temples and synagogues across the land are seizing the environment as a top-priority concern."[30] One ecumenical effort was the National Religious Partnership for the Environment, a coalition of the United States Catholic Conference, the National Council of Churches of Christ, the Evangelical Environmental Network, and the Coalition on the Environment and Jewish Life. The goal of the coalition, said Paul Gorman, its executive director, is "to weave throughout all areas of the faith life a distinctively religious contribution to the cause of environmental sustainability and justice in such a way as to contribute to the scope of vision, moral perspective, breadth of constituency, and endurance of the struggle for all efforts to protect the natural world." The environment is not seen as just another issue but as "an effort to integrate creation permanently into the fabric of religious life," Gorman explained, adding, "I think it is safe to say this has become an irreversible current." Indeed, "green theology" is springing up among groups as diverse as the Harvard Divinity School, which held a series of conferences on ecology and the world's religions in 1998; Jesus People Against Pollution in Mississippi; and the Redwood Rabbis in California.

But if the religious institutions are beginning to embrace an ecological faith, the environmentalists have not yet adequately reached out to the churches and synagogues, Gorman believes. He complained of what he called the "provincialism of the environmental movement" and averred that environmentalists could display greater respect for the teachings of the world's religions. "If you look at most environmental magazines and news magazines, you see a culture and civilization in which religion does not exist," he said.

The transcendental flame that kindled the earlier conservationists, whether it came from a formal deity, an Emersonian oversoul, or the deification of nature itself, has dwindled to a faint spark within the mainstream environmental organizations today. The deep ecologists have a point. The movement needs to reignite the transcendental fire, to rededicate itself to the beauty and sanctity of this planet. By no means can environmentalists depart from a rigorous scientific foundation for their argument, but science and spirit can be mutually reinforcing. As writer and conservationist Michael Frome insisted, "We need a community of faith—faith in nature, in humankind, and in each other. We need to set the highest goals possible and never abandon them." And as Paul Gorman said, "Spirituality, moral vision—that is the force that moves a social issue to an attribute of civilization."

The Environmentalism of Place

Squeezed by giant corporate agribusiness and by the competition of the global marketplace, the small family farms of Berkshire County in western Massachusetts, like those in so many other parts of the United

States, all but disappeared over the course of the 20th century. Lands that once produced vegetables and fruit and supported dairy herds and apple orchards, providing much of the food needed by the surrounding communities, are now second-home developments, retirement communities, shopping malls, or simply neglected fields filled with brush and scraggly second-growth forests. But one new agricultural operation, a small farm nestled among rounded hills outside the town of Great Barrington, is not only surviving but thriving.

The crops harvested by Sunways Farm are not exotic: greens, tomatoes, onions, radishes, strawberries, squash, pumpkins, potatoes. But the farm is unusual in one important way. It is part of a new movement called community-supported agriculture. Here is how it works. Each year before harvesting begins, the farmer, David Inglis, sells several hundred shares of his anticipated crop to residents of the surrounding area. In return, each shareholder may come to the farm once a week and pick up a shopping bag full of whatever Inglis and his helpers have harvested that day and laid out on shelves in the barn. Shareholders can also go out into the fields themselves and pick their own berries and flowers. For the farmer, the benefits are obvious. He is assured of a certain income. He can avoid the costs of marketing by having his customers come to him. The advance payments insure against loss of income due to acts of God.

The shareholders do not have a choice in what they get other than taking or leaving what is available each week. The price they pay for their food is probably somewhat higher than it would be in a super market, but the system rewards them in a number of ways that have little to do with money. The produce is always fresh, harvested that day, in fact. Food is grown without the use of pesticides or chemical fertilizers, so the shareholders need fear less for the health of their families. Many of them want to support the small, local farm because they know it will conserve open space, which is disappearing before a tidal wave of development. Perhaps most important, however, people in the community support the farm because it is a *place* they can come to, where they can see where and how their food is grown. It is a place in which they can walk around and see and touch and smell the food they will eat while it is still being grown. Instead of being the end point of a vast, impersonal, industrialized, chemicalized international food production and distribution system, Sunways shareholders can come and chat with the people who grow the food that will be on their table that day.

Community-supported agriculture is based on the teachings of British economist E. F. Schumacher, who urged a return to small-scale local and regional economies as a means of promoting prosperity, local self-reliance, and democracy; of preserving the environment and restoring civic responsibility.[31]

In his book *Community and the Politics of Place,* Daniel Kemmis, former mayor of Missoula, Montana, who now heads the Center for the Rocky Mountain West, states that "public life can only be reclaimed by understanding, and then practicing, its connection to real, identifiable places." He goes on to say that "the strengthening of political culture, the reclaiming of a vital and effective sense of what it is to be public, must take place and must be studied in the context of very specific places and of the people who struggle to live well in such places."[32] This is an admonition the environmentalists could well heed. Environmental organizations certainly are strong defenders of places such as national and state parks, wilderness areas, forests, wildlife refuges, grasslands, and wetlands. But they do not do so in the context of community, of places where people live and make their living.

The big national and international environmental groups in particular are appropriately vigilant defenders of places inhabited by wolves, bears, eagles, and whales. But they are less assiduous about helping residents of inner-city neighborhoods and rural communities inhabit healthy, attractive, and economically viable places. Although people care about parks and forests and other places they visit, they care fiercely about the places where they live, work, and raise their children. It is this fierce caring for place that the environmentalists only sporadically enlist in their cause, and this is one reason why the political culture of the nation gives them such shallow support.

Although the anti-pollution and resource conservation laws that the environmentalists help generate and enforce are designed to protect places and their residents everywhere in the country, they are often carried out in ways that disaffect local communities—an example is the effects of the Clean Air Act in coal country, where rules discouraging the combustion of coal throw miners out of work and communities into depression. This disaffection has provided fertile ground for the seeds of the "wise use" and property rights movements planted by extractive industries and right-wing ideologues. Daniel Kemmis contended that whole regions of the country, such as the Rocky Mountain West, have grown hostile to national environmental rules and programs that pay insufficient heed to the unique qualities and needs of those regions. That is why, he said, many western states are electing conservative, anti-environmental politicians and sending them to statehouses and to Washington, D.C.

Kemmis's solution to the problem is regional autonomy for the Rocky Mountain area (and presumably other regions) to allow it to make its own decisions on issues of land and resource use instead of having those decisions made by a national government. Noting that much western land is owned by the federal government, he said: "It is very hard for people outside the West to know how frustrating it is to be told over and over again that the land that surrounds you belongs to all of us and you don't have more control of it than the rest of us do. It is a very disempowering thing, and the environmental movement is more guilty of that disempowering than anyone else." He also asserted that national control has not provided the level of environmental protection most westerners want and that people living in "communities of place" would do a better job if given responsibility for managing their surrounding lands and resources. He pointed to what he called "a great explosion of collaborative efforts all along the political spectrum" at the local and community levels by those who want to establish local control of environmental decisions.

The idea of community residents having control of their environment is certainly in the democratic tradition and has obvious practical benefits. But there are also problems involved with transfer of authority from the national to the regional level or local level. These problems were illustrated by one collaborative effort to decide on environmental issues made in Quincy, California, where local citizens at fierce odds over logging in three nearby national forests came together to forge a solution. Local elected officials, local environmentalists, and representatives of the logging industry and other commercial interests met in the town library and were dubbed the Quincy Library Group. They came up with a plan that the local congressman, Wally Herger, a Republican unsympathetic to environmental causes, submitted as legislation. The Quincy compromise, which provided for additional road building and logging in the national forests, was hailed as a model of collaborative action replacing adversarial confrontation and was passed by both houses of Congress. But many environmentalist were outraged, saying that protection of national forests and American taxpayers had been sold out. An editorial in the *San Francisco Chronicle* called the Quincy plan "a sweetheart deal for the timber industry, wrapped in a quaint title and a veneer of consensus."[33] The Sierra Club's Michael McCloskey warned that the agreement established a dangerous precedent for national forests and for all environmental issues in which there is a national interest by excluding "nonlocal stakeholders" from the decision-making process. McCloskey also said that local collaborative efforts can often be controlled by industry, which has the power "to generate pressures on communities."[34]

So, where does that leave the environmentalism of place? It probably means that environmentalists, when seeking to influence law, policy, or enforcement or to carry out direct action, will have to consider very carefully what is the proper community of place in which to conduct their activities: the neighborhood, town, region, country, or globe or some combination of these. It also means that communities at all levels, when trying to reach solutions to their environmental–economic problems, should think more broadly about other constituencies with an interest in their decisions and, perhaps, should give them seats at the negotiating table when appropriate. The Quincy Library Group, for example, might have had smoother sailing if it had consulted with the national and regional environmental organizations. Daniel Kemmis insisted that communities of place are more important than communities of interest. That is no doubt so, but communities of interest, such as the national environmental community, are and will continue to be important and appropriate influences on public life.

One promising effort to grapple with the environmentalism of place is a set of principles propounded in 1998 by the Western Governors' Association called the Enlibra doctrine. *Enlibra* is a word coined to symbolize balance and stewardship. The principles call for collaborative efforts to provide "neighborhood solutions" to environmental problems but solutions that adhere to national standards and are based on objective scientific facts. An author of the doctrine, Oregon's Democratic governor, John Kitzhaber, described it as a way to create new tools for meeting environmental challenges and at the same time "build communities rather than disrupt them."[35]

National standards alone would not be sufficient to end the adversarial confrontations over environmental issues. National constituencies will have to have continued input into the process when there is a national interest at stake. New nongovernmental environmental institutions probably will have to be created to participate in such a decentralized decision-making process; the already overburdened mainstream groups could not cope with a system of environmental protection that is fueled by hundreds or thousands of "neighborhood solutions." But they can still help set and monitor the standards and help create and support the new institutions.

In the 20th century, many communities of place unraveled and social and civic institutions eroded in the face of suburbanization, globalization, mass mobility created by car ownership, the Interstate Highway System, and air travel; by corporate giantism and standardization of consumer products; by a volatile and shifting economic base; by junk television and the other destabilizing features of a trash culture; and by a rampant individualism that turned its back on the communal life and the commonweal to seek "self-fulfillment" or material success and gratification. Just as we set ourselves apart from nature, as Robert Bellah and his coauthors observed in *Habits of the Heart*, late-20th-century Americans "imagined ourselves a special creation, set apart from other humans."[36]

At the turn of the century, however, there is clearly a pervasive unease among the American public, a keen sense of a society gone astray. There also appears to be a strong yearning among many Americans to be part of a mutually caring community, to join in institutions that look after their members and protect and nourish their surroundings, to belong to and love a specific place on the earth.

The environmental movement can embrace this deep impulse as it prepares to take on the very difficult tasks it faces in the coming years. Concern about and love for place can make environmentalism a unifying force. Environmentalists can help create "a vision of what a community can be" and take part in turning that vision into the reality of a new society.

CHAPTER 5

The Business of America: Environmentalism and the Economy

"Capitalism, it is said, is a system wherein man exploits man," wrote sociologist Daniel Bell in *The End of Ideology*. "And communism—is vice versa."[1]

It might also be said that communism is a materialistic economic system that treats nature—the environment—as just another commodity to be ruthlessly exploited for maximum growth. And so is capitalism.

The history of the 20th century is in large part a narrative of the mortal struggle between antagonistic political and economic systems: totalitarianism versus democracy; communism against capitalism. By century's end, both democracy and market capitalism had emerged as victors, although in the case of democracy, the victory was far from absolute.

The proliferation of democratically elected governments around the world in the latter part of the century was a surprising, heartening triumph of the human spirit. A triumphal, unchallenged, unfettered market capitalism, however, is not an entirely unmitigated blessing. Nikita Khrushchev's prediction that communism would "bury" capitalism turned out to be a hollow boast, an ironic epitaph for the brutal Soviet experiment. Unfortunately, it may be capitalism that buries our society if it continues on its current path of grossly inequitable distribution of wealth, unsustainable economic growth, and devastation of the environmental systems that are indispensable to a livable, healthy, and secure habitat for humans and all other life.

Capitalism and the Environment

Like any economic system, capitalism is an artificial construct, one of the ways humans have devised to organize their society. No economic system, including capitalism, is immutable or infallible, as are, say, the laws of gravity and thermodynamics. The intellectual foundation of the capitalism still practiced today was laid by Adam Smith more than two hundred years ago in a world vastly different from the one we face at the beginning of the new millennium. Yet some adherents of market capitalism embrace it with the same kind of unreasoning fervor with which militant fundamentalists embrace their religion. As John Gray of the London School of Economics commented, "The contemporary cult of the free market is just as radical an exercise in social engineering as many exercises in economic planning tried in this century. Like other kinds of high modernism, it rests on a confident ignorance of the immensely complex workings of real societies."[2] It could be added that the free-market cult also rests on a willful indifference to the complex workings of the biological, chemical, and physical systems that support life on earth and the way those systems interact with the economy.

The market economy does some things very well. It is a powerful engine of production, an efficient distributor of its goods and services. It promotes innovation, it can create and increase wealth at a rapid rate, and for many Americans, it has raised living standards to levels undreamed of a century ago. At the end of the 20th century, the system appeared to be working beautifully for most of the American people. In January 1999, a beleaguered President Clinton, in what he noted was the last State of the Union Address of the millennium, boasted that the nation was experiencing the longest peacetime economic expansion in its history.

But for anyone who cared to look, the market economy was riddled with flaws. Our economy does poorly in distributing the wealth it creates. In the United States, the richest 1 percent of the population owns more of the wealth than the bottom 90 percent. The real wages of workers have been declining for twenty years.[3] Early in 1999, economist Lester C. Thurow noted that the median income of American families had not risen since the early 1970s, although American wives were working fifteen more hours per week outside the home. Almost 90 percent of the stock market gains had gone to the top 10 percent of families. Much of the new wealth had gone to "bigger-than-that-of-my-neighbor sport utility vehicles, 10,000-square-foot homes or other conspicuous consumption."[4] It cannot be said that the market economy has performed exceptionally in promoting social welfare and equity in the United States, not, certainly, if one looks at continued poverty and hunger; widespread drug, alcohol, and tobacco addiction; high, if declining, rates of crime, violence, and suicide; disintegration of families and communities; and other signs of social alienation. The economy is crucial to the quality of the human condition, but it is not the only crucial sector. Market economics does not deal well with issues of ethics or morality and, in fact, is largely blind to the broader world that encompasses human existence. As Frances Moore Lappé pointed out, the market economy responds to purchasing power, not to human needs.[5]

Free-market capitalism has also been conspicuously unable to preserve and protect our environment. That is why the landmark federal environmental laws were enacted and why command-and-control regulations—regulations that stipulate not only what must be done to protect the environment but also how it must be done—were instituted in the first place. Jonathan Adler, director of environmental studies at the Competitive Enterprise Institute, an institution that proselytizes for a laissez-faire approach to business, industry, and property and to the economy generally, argues that the environment is best protected when it is marketable private property. "Despoiling land reduces value—so private owners will safeguard it."[6] But all anyone has to do is look around to see how hollow that statement is. Try driving along the turnpike in northern

New Jersey to see in what fashion private property is protected by the industrial and commercial enterprises that line it. Look at the poisoned brownfields across America, the naked hills and sterile streams of Appalachia, the blasted, abandoned blocks of tenements in our inner cities; the barren strip malls of suburbia, the productive farmland disappearing under ticky-tacky housing developments, fields made hard and barren by excessive use of chemical fertilizers and pesticides. Private ownership may be a motivation for safeguarding property in some cases—but it depends on the kind of property, who owns it, and what the owner wants from it. And the free market fails miserably and absolutely to protect our common property—the air, the atmosphere, the health and safety of city streets and tenements, the lakes and rivers and oceans, the forests and watersheds, the abundance and diversity of life.

One problem is that the neoclassical economics of free-market capitalism simply does not account for environmental values. Market capitalism seeks to maximize the utility of resources—in practice, maximizing production, consumption, and the growth of capital. As economist A. C. Pigou noted early in the 20th century, the damages caused by economic activity—pollution and the depletion of natural resources—are considered to be "externalities" to the economy, although, as he pointed out, they impose real economic costs on society.[7] Or, as Herman Daly, a father of ecological economics, put it, "In the neoclassical view, the economy contains the ecosystem" rather than the other way around. This perspective regards nature as "just one more sector like agriculture or industry."[8] The marketplace is flexible and innovative because it responds to price signals, which in turn reflect producer costs and consumer demands. But because of its blindness with regard to the environment, it often fails to transmit and receive necessary signals. As futurist Hazel Henderson remarked, "Prices that do not incorporate the full spectrum of environmental and social costs can guide markets to unsustainable paths."[9] Brian Tokar put it more starkly: "A society that extols greed, acquisitiveness, and the unlimited accumulation of personal wealth simply cannot be expected to honor the integrity of life on earth."[10]

Illustrations abound of the market's failure to accurately price goods and services to reflect their full environmental, social, and even economic costs. For example, the price of gasoline at the pump was near an all-time low in real dollars toward the end of the 20th century. But this price did not include the ecological damage caused by finding, extracting, and transporting oil from increasingly remote places such as the deep ocean and the shores of the Arctic Ocean, nor the high cost of air pollution from combustion of oil and gasoline by industry and motor vehicles. These costs include huge health expenditures on heart and lung disease and other illnesses caused by smog and particles in the air, corroded buildings and other structures, and degraded water supplies and forests. And, of course, oil prices do not include the substantial portion of the huge military budget required to defend oil supplies in the volatile Middle East.

Even more problematic, the market hardly begins to price "natural capital," the stocks of natural resources provided by ecological systems such as forests and oceans on which life depends. These include raw materials such as lumber, fuel, and fodder; genetic resources; and services such as pollination, climate regulation, water filtration and storage, soil formation, and nutrient cycling, among many others. In 1997, economist Robert Costanza and colleagues estimated the value of seventeen services provided by ecological systems at an aggregate annual average of $33 trillion and possibly as much as $54 trillion. In contrast, the global gross national product—the value of all human-created goods and services—was $30 trillion per year. Costanza and colleagues found that because commercial markets do not adequately quantify the value produced by the earth's natural capital, it is often given little weight in policy decisions; they warned, "This neglect may ultimately compromise the sustainability of humans in the biosphere."[11]

Social critic Robert Kuttner wrote in 1998, "Half a century ago, the democracies of the west, chastened by two world wars and a depression, by the brutalities of pure capitalism and the menace of communism, concluded that a market economy needed to be tamed and domesticated and coexist with a decent stable and just society." But, he added, "the stagnation of the 1970s, the resurgence of organized business as a political force, and finally the collapse of communism, revived an almost lunatic credulity in pure markets and a messianic urge to spread them worldwide."[12]

The purveyors of this "lunatic credulity" gained formidable political power in the latter part of the century as right-wing ideologues took control of the Republican Party, won the White House during the Reagan administration, and then took control of the 104th and 105th Congresses. One of the chief uses they made of that power was in an effort to ease or remove governmental regulation, particularly environmental regulation, as an obstacle to the unconstrained functioning of the marketplace. Although that effort was only partially successful because of the American people's continued concern about the state of the environment, Reaganism and the "Contract with America" Republican conservatives did provoke widespread distrust of government and gain adherents to the free-market religion.

Some of those adherents are environmentalists. Indeed, the "third wave" of environmentalism is in large part a shift away from reliance on governmental enforcement to an increased reliance on market mechanisms and incentives to persuade corporations and other institutions to do the right thing in their own self-interest. The plan to allow power companies to sell sulfur emissions devised by the Environmental Defense Fund and incorporated in the Clean Air Act Amendments of 1990 is the most widely recognized example of this approach. And the scheme has produced some real benefits. As a result of the buying and selling of marketable emissions, the cost of reducing sulfur pollution from power plants, a chief cause of acid rain and urban smog, has turned out to be only a fraction of what industry said it would be during the debate over the amendments. Of course, industry almost invariably overestimates the cost of cleaning up the environment in its efforts to avoid doing so. But as Fred Krupp, executive director of the Environmental Defense Fund, noted, the emissions-trading plan has made many formerly hostile utilities willing adherents to the Clean Air Act amendments because it is in their economic interest to be so.

The sulfur emissions program, however, also demonstrates the limitations of a reliance on market forces. Although it has produced an overall nationwide reduction in emissions of sulfur from power plants, it has not reduced the sulfur pollution significantly in many areas that need it most. Thomas C. Jorling, former commissioner of the New York State Department of Environmental Conservation and later a senior executive of the International Paper Company, said: "The idea that the sulfur program is working beautifully is bullshit. It is working if you are an economist. But if you are in the Adirondacks it is not working well."

What is happening, he explained, is that pollution credits were sold by utilities in clean-air areas to plants in the Midwest, and these plants' emissions are being carried by prevailing westerly winds and continue to pollute the Adirondacks and the rest of the Northeast, causing acidified lakes and dying trees. In 1998, the Long Island Lighting Company, one of the country's most aggressive marketers of pollution credits, had to be persuaded to sign an agreement with the office of New York's governor, George Pataki, to keep its emissions credits away from out-of-state polluters whose emissions would harm the state.[13]

Patrick Parenteau, former director of the Vermont Law School's Environmental Law Center and former official of the Environmental Protection Agency, thinks that reliance on market-based incentives such as sulfur-trading emissions is "simplistic and naive and as damaging as evil intent. Corporations do what they do for a very clear and understandable and even legal set of necessities. They have to be forced to do the right thing through legislation, litigation, and the electoral process, and so they don't have any other choice."

There is a middle ground. The market is an enormously powerful instrument, and there undoubtedly are many areas where it can be enlisted in the service of environmental protection. But the world is far too complex to be governed by any single system of belief, even one as powerful and flexible as the market. Environmentalists will continue to need other tools, including the law, government, public opinion, and direct action of many kinds, to transform the marketplace into a reasonably friendly ecological milieu in the 21st century. An environmental agenda prepared in 1997 by the National Academy of Public Administration stated that the nation should attempt to the maximum extent possible to employ the "invisible hand" of the market to achieve environmental goals. However, it added that "paradoxically, a combination of market forces and public actions can help the nation" achieve those goals.[14] The carrot works best when the stick is an imminent option.

Environmentalism and Economic Growth

The most frequent argument heard from foes of measures to reduce or eliminate pollution and preserve natural resources is that these efforts cost too much money, absorb investment capital, slow economic growth, and result in the loss of jobs. This argument fueled the political debate that raged over environmental protection for most of the second half of the 20th century. It is most frequently advanced by corporations, economic entities that must bear a large share of the costs of correcting environmental abuses, and by ideologues who oppose virtually any requirement for private expenditure in the name of a public good. Environmentalism itself is often seen as anti-growth, an attitude, according to Walter Rosenbaum, that "places environmentalists on a collision course with dominant American values."[15]

The facts, however, do not bear out this view of the economic effects of environmental protection. In the mid-1990s, about $150 billion per year was spent in complying with environmental regulations in the United States. This was about one-third of the national expenditures for defense and about one-third or less of expenditures on medical care. It was one-half of what the nation spends on clothing and shoes. And although the costs of some environmental programs, such as the Superfund Program for cleaning up hazardous wastes, are climbing sharply, the growth of environmental expenditures declined from an annual rate of 6 to 8 percent in the 1980s to about 3 percent in the 1990s.[16] One reason for this decline is that the political climate discouraged new environmental laws and regulations. In any case, the total spent on the environment is trivial in the context of a $7 trillion gross national product (GNP)—although, as we shall see, the GNP is regarded by many as a doubtful measure of growth and prosperity. For one thing, it fails to measure as deficits many economic activities that degrade the environment, such as the health costs imposed by polluted air or the depletion of natural resources such as forests. These things are regarded as "externalities"—external to neoclassical economics.

As for jobs, the Bureau of Labor Statistics found that only 0.1 percent of layoffs in the United States could be attributed to causes related to environmental protection. The other 99.9 percent of layoffs were due to other causes, such as automation, corporate mergers, and exportation of jobs to low-wage countries. Some studies, in fact, have found a correlation between economic growth and the existence and enforcement of strong environmental laws. The discipline imposed by these laws breeds efficiency in the workplace and the marketplace and opens new economic opportunities, such as in environmental control technologies.[17]

There is no free ride, of course, to a clean, healthy, pleasant, and productive environment. Environmental protection comes at a cost. But that cost may not be even as high as the raw data indicate. A study conducted by Resources for the Future found that the cost of environmental protection may actually be *overstated* because efficiency and productivity gains associated with changes made in production technology and processes to meet environmental standards can substantially lower production costs. The study, which evaluated large industrial plants, found that each dollar spent in complying with environmental regulations may, on average, result in only thirteen cents of additional costs.[18] Such estimates do not, of course, include the net gain to society when environmental benefits are included in the equation. A dollar spent by an industrial plant, for example, could produce more than a dollar's worth of reductions in health expenditures in its community and, by reducing days lost to ill health, add to the productivity of workers in that community. And when measured on an industry-by-industry basis, less than one full-time job is lost for every million dollars spent in meeting the requirements of environmental regulation.[19]

This kind of data—and there is a substantial and increasing body of such information—should effectively counter the arguments of industry and free-market ideologues that environmental rules batter the economy and throw Americans out of work. At the end of the century, after thirty years or more of federal regulation and other intensive efforts to reduce pollution, safeguard health, and protect our resources, the economy was robust, inflation was low, and unemployment was lower than it had been in decades. Nevertheless, many corporations, their trade associations, and assorted other anti-environmentalists continue to sound the same old off-key alarm about the economy. Of late, proposals to reduce carbon emissions to mitigate the effects of global warming have had the fossil fuel industry and its allies loudly crying wolf about supposed economic havoc. A statement signed in 1997 by 2,500 distinguished economists, including eight Nobel laureates, affirmed that carbon dioxide and other greenhouse gases could be cut in ways that would not reduce American living standards, could increase productivity, and would produce benefits that clearly outweighed the costs. But industry, particularly the fossil fuel sector, following what it perceived as its own self-interest, continued to employ the argument of economic catastrophe in its campaign to block measures to lessen the threat of climate change.

One lesson here is that environmentalists need to do a much better job of accumulating economic evidence and, even more vital, bringing that information to the attention of the public in ways that move the public to respond. Countering the heavily funded disinformation campaigns on the economic consequences of environmental protection mounted by corporations, trade associations, and conservative think tanks will be difficult, but it must be done.

Although the macroeconomic effects of environmentalism may be trivial, environmentalists cannot ignore the economic costs of their activities on workers and communities. Statistics about the low average effect on employment and economic growth are little comfort to individual laid-off workers and their families or people who live in depressed communities. The economy of Oregon, for example, may be booming as new high-tech industries move in, but over the short run—and the short run is when grocery bills and rent must be paid—that does not help loggers and sawmill workers who are out of work because of reduced cutting in the national forests to protect the northern spotted owl.

The leaders of a growing number of environmental organizations are beginning to understand that they must do a full cost accounting of the economic effects of their work. If the economic climate does not support their goals, those goals will not be achieved. The Rocky Mountain Institute, for example, has been working with timber-dependent communities in Oregon to help them create diversified, sustainable economies. In Oregon's Applegate Valley, the institute is helping to expand the area's young organic agricultural industry as well as helping to develop an environmentally sensitive timber sale.[20] In the 21st century, environmentalists will have to consider themselves as responsible for workers and communities as they are for owls and clean air.

Another lesson is that environmentalists and others responsible for protecting the environment in both the private and public sectors have to find more efficient and economical ways to achieve their goals. Harvard economist Dale Jorgenson pointed out: "We have had a lot of improvement in the environment. And we have paid a lot for it. The question is, have we paid too much for what we are getting? And if we are able to do it more cheaply, could we get even more improvement in environmental quality? And the answer is yes."

But is economic growth good or bad for the environment? Certainly the environment suffers the worst insults in such places as debilitated inner cities and impoverished countries, where economics are depressed and people must scratch at their surroundings for the basics of life. Without economic growth, those countries and communities and their inhabitants are unfairly locked into their poverty and a habitat condemned to progressive degradation. They must eat the seed corn. Conversely, in rich countries and communities, where living standards are high, the economy affords surpluses that can be invested in preserving land and resources and replacing industrial technologies that degrade the land, air, and water.

Yet on the other hand, in a burgeoning, prospering economy, the exploitation of raw materials, the production and consumption of goods, and the disposal of waste increase rapidly, sometimes exponentially. Such growth shrinks resource reserves, fills waste sinks, reduces and degrades water supplies, pollutes the air, spreads dangerous substances, and changes the physical and biological systems that support life on the planet.

Economist Kenneth Arrow and colleagues wrote that "economic growth in itself is neither the problem nor the solution in dealing with the environment. It is neither a panacea for improving environmental quality nor does it have an inherently negative impact on the environment. *What is important is the content of economic growth* [emphasis added]. It is better to concentrate on internalizing external environmental effects and thereby protecting the resilience of ecological systems. . . . [T]his is not an argument against economic growth per se but against the presumption that economic growth will automatically resolve the problem of environmental degradation and that the economic growth is automatically environmentally sustainable. Even low income countries can and should improve efficiency and hence welfare by internalizing environmental externalities and, even high income countries can and should restrict consumption and production so as to protect the resilience of the natural systems on which they depend."[21]

In recent years, certainly since the 1992 Earth Summit in Rio de Janeiro, the economic mantra of most environmentalists, to which most governments also pay lip service, has been sustainable development. The most widely used definition of sustainable development is that presented in *Our Common Future*, the 1987 report of the World Commission on Environment and Development: "development that meets the needs of the present without compromising the ability of future generations to meet their own needs."[22] The report goes on to say that economic growth will be needed if the poor of the world are to achieve a decent standard of living. But development does not necessarily mean growth. Herman Daly asserted that "the term 'sustainable growth' when applied to the economy is a bad oxymoron." He added that "it is impossible for the world economy to grow its way out of poverty and environmental degradation" because the economy "is an open subsystem of the earth ecosystem, which is finite, non-growing and materially closed."[23] Environmentally sustainable economic growth, however, is probably achievable. What is important is the *content and use* of economic growth. We can have economic growth for purposes different from the present purposes: growth that is based on careful husbanding of resources, on more benign technologies; growth that does not exceed the carrying capacity of the planet's resources and systems; and growth that more equitably distributes wealth than is the case today. It would be growth designed not to increase our consumption of ephemera but to improve the quality of our lives, preserve the beauty and salubrity of our surroundings, and secure the future of our posterity. Early in 1999, Vice President Al Gore introduced a plan for "smart growth," which called for community economic development that includes preserving green space, protecting water supplies and other natural capital, reclaiming polluted industrial sites, and building public transportation and other infrastructure to reduce dependence on the automobile.[24]

Although Gore presented the concept as a way of preserving the landscape and providing lifestyle amenities, smart growth would replace a pattern of economic growth pursued by real estate developers, commercial enterprises, extractive industries, and builders of industrial facilities that seek the cheapest land and the quickest access to natural resources such as trees and water in order to maximize their return on investment.

This smart growth agenda only touches the surface of what would be required for an economic system to sustain the natural world and its human inhabitants into the distant future, but such a proposal, introduced at a high level of government, at least puts the notion that we need a different kind of growth, a different kind of economy, into the mainstream of national affairs.

There are many obstacles to making the difficult transformation to that kind of economy, including our definitions of what constitutes growth and the ways we measure growth. Most of us, including economists, think of growth as more production, more consumption, and more accumulation of wealth. Our chief measure of growth is the gross national product, which counts the aggregate output of goods and services even if those goods and services are harmful to the environment, to public health, and to human happiness. Herman Daly contended: "Rising GNP is not making us richer. In fact, it is making us poorer if further GNP growth increases environmental and social costs faster than it increases benefits." Although the federal government's Office of Management and Budget and others have formulated economic indices that measure components of human welfare and quality of life, we are still more likely to refer to the GNP and the Dow Jones Industrial Average when we talk about the state of the economy.

A major task facing environmentalists in the 21st century, therefore, is to participate in formulating new definitions of what constitutes just, proper, and rational economic growth and to help plan the transition to an economic system that preserves and enhances the environment while fairly dividing the blessings of a thriving economy. Environmentalists must help build an economic system patterned after nature's economy, one that grows not by continual production and consumption but by constant self-renewal.

Corporations and Environmentalism

Corporations are arguably the most powerful, influential institutions in the United States and throughout the world. Certainly they command overwhelming economic power. The assets of the 500 largest U.S. corporations were more than $2 trillion by the beginning of the 1990s[25] and were approaching $3 trillion by the end of the 1990s. This wealth was greater than that of all but two or three nation-states. Indeed, the economic size of single corporations such as the Exxon Corporation and the General Motors Corporation was greater than that of most of the nations of the world. The world's 500 largest corporations control 25 percent of the world's economic output but employ just 0.05 percent of the world's population.[26] The trend of mergers and acquisitions in the latter part of the 20th century concentrated ever more economic power into fewer and fewer megacorporations.

Corporations large, medium, and small wield their economic power to influence—and in some cases control—virtually every aspect of our lives. Through advertising expenditures, which are some $200 billion per year in the United States, and vast, skillful marketing techniques and networks, they dictate our consumer choices and, in many ways, our lifestyles. With billions spent on public relations, they play a major role in creating consensus not just on market issues but also on social issues. Disinformation campaigns by corporations and their agents, for example, are often used to thwart efforts to protect the environment. Through ownership and concentration of the communication media, they filter information to the public and thereby influence the way we think and the content of our culture. Toward the end of the 20th century, those who controlled the media were increasingly blurring the line between information and entertainment.

Almost needless to say, the lives of workers and the fates of communities are decided in corporate boardrooms.

The economic might of corporations also gives them political influence far exceeding that of any individual and of other institutions in American life. The role of money in politics is discussed in the next chapter, but it should be noted here that with their unstinting expenditures on lobbying, either directly or through their trade associations, the flood of money they direct to political candidates and parties, their ability to communicate with the public through advertising and public relations, and the media they control, corporations—business and industry—have become the dominant political force in our increasingly compromised democracy. This is a fact of life at all levels of politics, from small communities to Washington, D.C. Even small businesses wield undue power through their chambers of commerce, their Rotary clubs, and other institutions in which they pool their economic and political clout.

Corporations also have achieved legal standing and powers far more formidable than those of individuals or other institutions in our society. How they did so is a long and complex story, but a crucial bridge was crossed in 1886 when the United States Supreme Court ruled in *Santa Clara County v. Southern Pacific* that a private corporation was a natural person entitled to all the protections of the Bill of Rights and the Fourteenth Amendment.[27] Ever since, courts and governments at all levels have been expanding the protections, rights, and privileges afforded corporations, including grants of limited liability to corporate shareholders and grants of corporate charters in perpetuity. Such privileges do not place corporations above the law, but they do make them much more difficult for the law to reach and also insulate them from strict accountability for their actions.

The globalization of business is putting corporations further beyond the reach of governmental and democratic control. Regulatory agencies set up by governments to constrain corporations have far too few resources to match those of the corporations and can deal only with their grossest abuses.

The 1886 court ruling is bad fiction. Corporations are not people. As Stephen Viederman, president of the Jesse Smith Noyes Foundation, noted, corporations by their very nature have no commitment to community and place, to democracy, to the long term, to the participation of shareholders or other stakeholders in their decision making, to ending economic inequity or assuming responsibility for the costs of their actions to the public.[28] Yet when combined with their economic and political resources, the standing of corporations as individuals protected by the Bill of Rights makes them almost untouchable oligarchs in our society.

With their extractive use of raw materials, their synthesis of new substances, their industrial processes, the hazardous substances and other wastes they spew into the air, water, and soil, their stimulation of consumption, their power to dictate consumer choices, and their relentless drive to expand markets, corporations in the aggregate are the most important determinants of environmental quality in the United States and in other countries with developed economies. After decades of regulation and public pressure, many corporations have taken measures, some of them far reaching, to reduce their adverse environmental effects. Others continue to impose a particularly high ecological toll, especially those in the energy, automotive, chemical, forest products, mining, agribusiness, and real estate development sectors.

It is the cumulative effect of industrial and commercial activity that compromises our prospect for living in a healthy, thriving, spacious, and appealing habitat, and, indeed, that threatens the course of evolution and the future of life on earth. William Ophuls and A. Stephen Boyan Jr. argued that "because the basic premises of modern industrial civilization are anti-ecological, all its values, practices and institutions are grossly maladapted to the emerging age of scarcity."[29]

Clearly, then, if we are to achieve ecological sanity in the 21st century, the values and practices of our corporate institutions will have to change dramatically.

Some astute observers believe that corporations are already changing in the face of ecological and economic reality and that in coming years they will be not the problem but the solution to our environmental dilemma. Peter Raven, director of the Missouri Botanical Garden and frequent commentator on ecological issues, believes that "government is poorly equipped to lead us into the sustainable economy we so desperately need in the 21st century. Business, on the other hand, cannot afford to dupe itself on environmental issues and sustainability issues because, if it does, it will be unable to remain profitable." Noting that that the British Petroleum Company's chief executive officer (CEO), John Browne, broke ranks with executives of other major oil companies to endorse the Kyoto Protocol on climate change and to announce that his company would invest heavily in renewable sources of energy, Raven said, "You would have to be a moron not to see that BP will be in far better shape in ten years than the companies who say, 'We don't believe in that stuff.'" Designer Victor Papanek noted that some industrialists, mostly in Europe and Japan, "have recognized the current environmental and ecological hazards for what they really are: vast new challenges for humankind that must be solved, and *vast possibilities for future earnings, since few governments or industrial powers have yet taken these threats seriously*" (emphasis in original).[30]

A few corporations in the United States, and more in western

Europe, began subscribing in the 1990s to a new industrial philosophy dubbed "The Natural Step." The approach was formulated by Karl-Henrik Robèrt, who had headed Sweden's leading cancer institute, as a way to cut through the endless scientific and economic debates about environmental issues to establish a cyclical economy based on four principles. The four principles, or steps, can be summarized as follows: (1) materials from the earth's crust, such as coal and oil, should not be permitted to accumulate in nature; (2) persistent substances produced by human society, such as DDT and chlorofluorocarbons, should not be allowed to accumulate in nature; (3) there should be no systematic deterioration of the earth's natural cycles and biological diversity—no habitat destruction or fishery depletion; (4) resources must be used fairly and efficiently to meet human needs.[31]

The biggest American company to give at least lip service to The Natural Step is the Monsanto Company. Until the 1990s one of the world's major chemical producers (and producers of hazardous pollution), Monsanto spun off its chemical operations except for agricultural chemicals and is now what its CEO, Robert Shapiro, calls a "biology company" specializing in genetic technologies. Shapiro joined the company in 1990. He found the staff at Monsanto committed to lessening the company's damage to the environment by reducing its hazardous emissions by 90 percent. But the company was looking for new ways to create value, to move ahead into the next century. Monsanto determined, Shapiro explained, that "in addition to reducing the damage we do, we can be useful as a company with a lot of scientific, technical, and financial resources. In particular, we could devise technologies that would help move the world to sustainability, and do it in the context of helping with the economic development of the poorest people of the world. . . . I believe there are huge value-creating opportunities in providing what the world wants, and that includes economic development and sustainability." By its commitment to sustainable operations and products, Shapiro added, the company is able to attract highly qualified, committed professionals. "There is enormous economic value [in] having a bunch of people who care passionately about what they are doing," he said.

Not everyone is convinced of the sincerity or, at least, the efficacy of Monsanto's new sustainability direction. Environmentalists and other critics accuse the company of plunging ahead with radical genetic technologies before their potential consequences are fully understood. They note that one of the company's major products is genetically altered soybeans that can withstand large doses of Monsanto's Roundup herbicide, which environmentalists have been attacking for years. Poor Mr. Shapiro was the victim of a direct hit in the face with a tofu custard pie wielded by an activist who accused Monsanto of "greenwashing" itself as an eco-friendly corporation while continuing to market hazardous substances.[32]

Given the radical nature of genetic technologies and their potential for disaster, the jury is still out on their long-term consequences. However, because their potential benefits for the world are huge, environmentalists ought not to be in a position of knee-jerk rejection. What they should do is insist adamantly on rules to prohibit such technologies from being placed into the natural environment without exhaustive testing for safety and then only with stringent safeguards.

If corporations are to be a major part of the solution to the environmental crisis, they still have a long way to go. One company in the vanguard of cyclical production is Atlanta-based Interface, Inc., which leases industrial and commercial carpets until they need replacement and then takes them back and recycles their component materials. But as Paul Hawken, a founder of Smith & Hawken Inc., and a leading proponent of ecological industrialism, noted, Interface is pretty much alone among larger companies. "You could say Ray Anderson [Interface's CEO] is the exception that proves the rule, or you could say he is a harbinger," he said. "I prefer the latter." Although corporations are more committed to improving their environmental performance than the public realizes, Hawken said, he believes that many corporations do not "understand the problem at its deepest level. Too many still think in terms of industrial toilet training when the real problem is overall industrial metabolism. . . . Most businesses still think they are in the materials business when they are in the service business. Once that shift in thinking occurs, great things can happen, and you can get the Interface phenomenon, where innovation drives investment, drives sales, lowers throughputs, raises morale, increases margins, incites competition. You get real revolutionary change in that it is a cycle that feeds itself."

David Buzzelli, former vice president and corporate director for environment, health, and safety of the Dow Chemical Company and former cochair of the President's Council on Sustainable Development, contended that corporations "have made a terrific amount of progress on their emissions and on the safety of their operations. The next revolution is with the products and services we produce and the footprint those products and services make on the environment. I really don't think we will get sustainable development until business takes responsibility for these issues. We need to translate sustainable development and environmental concerns into some kind of economic system. We haven't done that."

At century's end, it was difficult to tell whether corporate America was in a state of transition or a state of confusion with regard to the environment. Bruce Piasecki, director of environmental management programs at Rensselaer Polytechnic Institute, said he thinks that corporate management is caught up in three "competing instincts." One of them is compliance with the law—no more than that, and less if possible. A second instinct is to meet public expectations for good environmental behavior through public relations campaigns, issues management, and shareholder relations strategies but not to change production methods, products, or marketing. The third instinct is to recognize environmental needs as a business opportunity and to change the product line to take advantage of that opportunity. Many companies respond to all three of these instincts, which often are in conflict with one another, Piasecki said. He cited General Electric Company (GE) as a prime example of a conflicted company. GE, he noted, has been seeking to develop products that could profit by the growing global ecological consciousness, but it is still in a state of confrontation and denial over the polychlorinated biphenyls (PCBs) its operations have spilled into the environment. The company has spent years and millions of dollars denying that PCBs are a health threat and resisting demands that it clean up contamination in such places as the Hudson River and Pittsfield, Massachusetts.

Unfortunately, the structure of the marketplace and of corporations themselves continues to encourage many, if not most, companies to follow their worst instincts about the environment and other social welfare problems. Because of corporations' endless need for capital for expansion and acquisition, their policies are often in thrall to Wall Street, which rewards short-term profitability above all else. "Wall Street sees environment as irrelevant, or bad news waiting to happen," said Matthew J. Kiernan, executive managing director of Innovest Strategic Value Advisors, Inc., a firm specializing in finance and the environment.[33] The discount rate that corporations use to write off costs also contributes to a short-term horizon that places little value on sustaining the environment and its resources. In effect, the discount rate writes off the future—valuable ecological assets today are worth zero thirty years from now under this system. As Herman Daly noted, corporations ignore social and environmental costs in the name of competitiveness. Executive salaries, value of stock options, and progress up the corporate ladder are also tied to a company's short-term financial performance, so for many corporate executives, self-interest tends to drown out social responsibility. Since World War II, the top managers of many corporations have often come from the corporations' marketing, financial, and legal departments rather than from their engineering and production departments, making those companies look to financial and marketing techniques for profitability rather than to innovative design and production. Business schools generally are doing an inadequate job of teaching their students about the gravity of environmental issues and training them for the tasks of environmental management. The participation of corporations in trade associations, for the often dubious benefit of industry solidarity, frequently drags many of them down to the lowest common denominator on environmental policy and performance. Presenting a united front may enable an industry to concentrate its lobbying, publicity, and legal resources, but it often can promote the interests of some companies at the expense of others and can present a negative public image of the industry as a whole. An example was the effort of the Chemical Manufacturers Association in 1980 to weaken the pending Superfund law over the objections of E. I. du Pont de Nemours and Company and several other chemical corporations.

In short, it appears that corporate America has some way to go before it is ready to lead the nation to an environmentally sustainable economy. As Joanna Underwood, president of INFORM, Inc., an independent research organization that monitors industry's environmental performance, observed, environmentalists will have to "keep pushing."

Environmentalists and the Economy

Where to push? One place is where the environmental movement has been pushing all along: insisting on high governmental standards and tough, enforceable regulations. Even if market forces and enlightened self-interest are leading corporate America toward ecological responsibility—and that remains an open question—the pace is still glacial. To encourage more rapid progress, there still needs to be a stick with which to beat recalcitrant polluters, resource depleters, and others who exploit the natural world. But the rules need to be more flexible than those adopted in the first flush of environmental activism. The rules should require that corporations, and municipalities, too, for that matter, adhere to strict federal and state standards, but they should permit them to do so by the most efficient means. Self-imposed industry standards such as ISO 14000 voluntary international safety guidelines should be encouraged but only if they are found to be strong enough to protect the environment and workers' health and safety and are actually put into practice by industry. Whatever their flaws, the market and its corporate components are highly efficient in reducing risk and cost to themselves and in maximizing profitability. Regulations that allow them to do that are more likely to be honored—as long as the alternatives include costly penalties.

Environmentalists need to continue to insist on full disclosure of the effects of economic activity on the environment and need to seek to expand the sources of information available to the public, the media, and the government. The Environmental Protection Agency's Toxics Release Inventory (TRI), a database containing information on the release and other waste management activities of toxic chemicals by facilities that manufacture, process, or otherwise use them, was developed in response to pushing from INFORM, Inc. and other environmental groups. The database has given citizens valuable information about hazardous industrial substances that could affect their health. It has also forced corporations to examine their own polluting activities more directly and to act to reduce such pollution. The TRI needs to be expanded to include the entire spectrum of potentially hazardous substances placed into the environment.

A full public accounting is also needed of the activities of corporations and other economic entities that may harm public or worker health and cause ecological damage. A growing number of corporations are now conducting annual self-audits of their environmental, health, and safety performance and releasing the results to the public as well as to their own shareholders. These audits—some of them, at any rate—appear to be reasonably honest and thorough. But the public should not have to depend on the goodwill and honesty of corporate managers or their desire to project a "green" image for dependable audits of their behavior. As soon as it becomes politically feasible, the environmental community should press for legislation, state by state if necessary, to require corporations to have legally binding annual environmental audits by independent third parties just as today they have their financial books audited. Among other things, such audits would include life cycle analysis and full cost accounting of all operations, products, and services—that is, accounting for all the environmental costs of their business, from acquisition of materials to final disposal or recycling. Like the TRI, such audits could give the corporations themselves a better idea of their environmental, health, and safety problems and potential solutions as well as provide information on which individuals, communities, governments, and environmentalists could act.

Certainly environmentalists should continue to seek ways to prod the market economy to perform in environmentally friendly ways. For example, current efforts by Friends of the Earth, in alliance with some fiscal conservatives, to rid governmental budgets of direct and indirect subsidies to industries such as mining, agribusiness, timber, highway construction, and energy extraction ought to get a higher priority and more support than they now do. On the other hand, the environmental community ought to press for governmental support of innovative industrial or commercial activities that promote sustainability, such as photovoltaic technologies and public transportation, through subsidies and purchasing agreements.

Environmentalists ought to campaign for institutions and individuals to invest in the securities of companies whose activities help, not hurt, the environment and sustainable economic growth. And they need to make sure their own endowments are invested in environmentally and socially responsible companies. Such an investment strategy should not be a difficult sell. Mounting evidence shows that companies with good environmental performance do well on Wall Street. An exhaustive study by The Alliance for Environmental Innovation, a nonprofit organization in Boston, found that companies that outdo their competitors in environmental performance also outperform them in the stock market. Conversely, the study found no evidence of a negative relationship between environmental and financial performance.[34] Environmental organizations, foundations, and their allies can also aggressively use their positions as shareholders to demand reforms in corporate behavior. As Stephen Viederman asserted, "Institutional investors cannot subscribe to the ostrich approach, denying responsibility for the impacts of their investments. If they take their responsibilities seriously, changes for the better can occur."[35]

One of the potentially most effective approaches to changing market behavior in ways that will benefit the environment is basic reform of the nation's tax structure. Proposals for reform have been before the public for some time now. In essence, they would reduce taxes on things that are economically and socially valuable, most specifically wages and capital that are now the basis of government revenues, and replace them with taxes on things that harm the environment and reduce the potential for long-term stability. These taxes would be imposed on the depletion of resources such as minerals, timber, and fisheries and on sources of pollution such as combustion of petroleum. The proposals envision a revenue-neutral tax, meaning that the total tax burden on individuals or companies would not go up or down. As a Worldwatch Institute report noted, "Using tax policy to steer the economy in an environmentally sustainable direction takes advantage of the inherent efficiency of the market."[36] Harvard's Dale Jorgenson commented that "tax proposals have never flown in this country because environmentalists have never bought into the idea." Well, some have, but not enough of them and not at a sufficiently high decibel level.

The actions suggested here so far would be useful in helping mend flaws in the economic system that permit destruction of the environment and assaults on public health. But with the exception of tax reform, they add up to little more than adjusting the accessories of the vast and complex machinery of our economy. What is really needed is for those in the environmental movement, in alliance with other social activists, to obtain enough economic *power* to compel economic behavior that protects and is in harmony with nature and meets the needs of all citizens. One way to achieve a more level playing field would be for our society to reclaim some of the power it has yielded to the corporations.

Richard L. Grossman, codirector of the Program on Corporations, Law & Democracy, and his colleagues have proposed a campaign to require that states revoke existing corporate charters and in their place issue charters that limit corporations' constitutional and legal powers and make them subordinate to the democratic process and to public sovereignty. The new charters would have written into them requirements that the corporation serve the general public good, including prohibitions against harming the environment.

Grossman believes that such a major restructuring of corporations and their powers is achievable. He pointed out that in the early years of the Republic, corporations were granted charters only for a specific period of time and a specific purpose and that states and local governments were able to exercise close oversight of their activities. He noted that that the issue of dangerously excessive corporate power has been raised in recent times at the highest judicial levels and cited a dissenting opinion by United States Supreme Court justices Byron R. White, William J. Brennan Jr., and Thurgood Marshall in the 1978 case *First National Bank of Boston v. Bellotti:* "It has long been recognized, however, that the special status of corporations has placed them in a position to control vast amounts of economic power which may, if not regulated, dominate not only the economy but also the very heart of our democracy, the electoral process. . . . *The State need not permit its own creation to consume it*" (emphasis added).[37] Environmentalists, Grossman said, can join with other citizens, including workers and community activists, to reclaim from the corporations sovereignty over the nation's political and economic affairs. Some communities are already beginning to act, Grossman noted. He offered as an example the city of Arcata, California, which has held meetings and passed a ballot initiative to explore its authority to control the activities of the MAXXAM Corporation, a conglomerate seeking to clear-cut the Headwaters Forest in Humboldt and Mendocino Counties, among the last remaining large stands of giant redwood trees in the world.

Grossman conceded that there is no quick and easy way for citizens to take back the constitutional privileges and legal status of the great corporations. It will be, he said, a long-drawn-out effort much like the years of litigation that led up to the Supreme Court's decision in *Brown v. Board of Education,* which ended legal segregation of public schools in the United States. But it can happen.

Finally, and potentially most significant, is the recently dawning recognition among some environmentalists, as well as within the labor, community-based development, and civil rights movements, that the way to win economic power is to become capitalists.

Dan Swinney, executive director of the Midwest Center for Labor Research, noted that the economic goals of those striving for social justice—workers, the poor, minorities, environmentalists—used to be centered on redistributing wealth to get a bigger share for their own use. Decisions about the means and use of production were left to corporations and money markets. In recent years, however, decisions about the investment and distribution of capital have led to plant closures and layoffs, a decline in real wages, increased concentration of wealth, and deterioration of the productive capacity of the environment.

The solution, Swinney contended, is for communities, workers, and environmentalists to assume responsibility for investment in means of production and creation of wealth. The creation of wealth cannot be left, as in the past, to those who have disqualified themselves by doing so in ways that "poison the earth." This is where the whole notion of sustainable development emerges; this is where people see a new vision of how to create wealth. It is a vision, Swinney added, around which to organize a new progressive coalition that brings together businesses, labor, environmentalists, and communities: "This is the alternative vision replacing the burned-out vision of the 1960s—something worth fighting for, something that's practical and achievable."

The idea of environmentalists investing in the means of production and engaging in the creation of wealth to be used in building a sustainable economy may seem at first blush to be radical or utopian. In fact, it already has started to happen.

Perhaps the best-known, and certainly the biggest, capital project involving an environmental organization is the Bronx Community Paper Company, a paper-recycling plant capitalized at more than $600 million. The Natural Resources Defense Council (NRDC) has been developing the company in partnership with Banana Kelly, a community development organization in the South Bronx. NRDC put its own money into the project and leveraged substantially more from state and city governments and private businesses. The project is intended to confront a major environmental need, the recycling of the huge amount of wastepaper created by businesses in New York City—the plant will convert some 330 million metric tons of paper per year—while at the same time addressing the needs to create jobs and alleviate poverty, to build infrastructure, and to reclaim environmentally debilitated land in one of the country's most distressed communities.

Allen Hershkowitz, a senior scientist for NRDC who has devoted his life, energy, and spirit to the Bronx project for more than a decade, is convinced that failure to participate in the industrial process is the "weak link" of the environmental movement. "Investment," he said, "is not only the key to sustainability; it is the key to alleviating poverty as well. The major problems to which I have devoted my life, the environment and poverty, are both remedied by targeted investments." A component of the environmental movement, he declared, "should engage in risk-taking entrepreneurial enterprises. We should do that in tandem with our legislative and enforcement work. Of course, we have to be on guard to make sure there are no conflicts of interest. But the environmental movement has to be involved in development in a much more substantial way. The movement is skewed too heavily to legal action and lobbying. We should be balanced by troops of M.B.A.s and Ph.D.s. We should be infiltrating Merrill Lynch, Goldman Sachs, Morgan Stanley and saying, 'Here is a good idea for an investment project.' The opportunity is there. The question is, will the movement be able to get its act together and take advantage of it?"

As of this writing, the Bronx paper project is still not off the ground. It has been beset by seemingly endless setbacks, including a lawsuit over use of the land, vicious fighting among competing community groups, and potential commercial partners who reneged on their commitments. Even if the project ultimately does not succeed, it can be seen as the beginning of a learning process for entrepreneurial environmentalism, a pathway to a new approach in the 21st century.

Meanwhile, there are a growing number of projects around the country in which capital investments are being made by nontraditional entrepreneurs in the service of environmental health and other social goals. The community development projects and the community-supported farm mentioned in the previous chapter are examples of entrepreneurialism in the public interest. In some areas of the country, including Berkshire County, Massachusetts, towns and villages are issuing their own local currencies to support local stores and market farms. The dollar, like other national currencies, basically reflects the nation's relative performance on the international market, not the performance of local economies. National currencies tend to flow from local economies to financial centers, where they are used for economic activity that does not necessarily benefit the local economy from which it flowed.

But local communities can use currencies for their own economic *and* ecological benefit. In Great Barrington, Massachusetts, for example, members of a nonprofit organization called SHARE helped two organic farms issue a currency (with a motto of "In Cabbage We Trust") that was redeemable in farm produce. SHARE also issued a loan to help build a barn for an ecologically responsible timber operator who hauled logs from the forest by horse instead of with machines to minimize adverse effects on the trees and undergrowth.[38] Food-buying cooperatives, community market gardens, workers' cooperatives that rely on sustainable resources, and even land trusts are further examples of entrepreneurial environmentalism.[39]

Suggestions that environmentalists own the means of production and participate in the creation of wealth undoubtedly will elicit cries of socialism from right-wing ideologues. But the dichotomy between socialism and capitalism is growing increasingly stale and irrelevant in the face of ecological and economic reality. Neither system, or, for that matter, any other rigid ideology, is likely to prove adequate for dealing with the onrushing economic, environmental, and social problems of the new century.

Members of the European Green parties like to assert that they are neither left nor right but out in front. Similarly, in the 21st century, our environment may flourish in an economy that is neither capitalist nor socialist but pragmatic, just, and sustainable. What we will need is what landscape architect Robert L. Thayer Jr. describes as a *"regenerative"* economy, an economy that meets the country's social, psychological, and ecological needs as well as its economic needs. In such an economy, he wrote, "both a worker's sense of self-worth and job satisfaction are regenerated along with the physical goods, lands, resources and social institutions involved. An economic structure can only be sustained if it regenerates order, stability, health, self worth, social structure, and ecosystem richness and viability."[40]

First, however, the environmental movement will require a reasonable and welcoming political climate if it is to change and strengthen our economy and help preserve our habitat and our civilization in the years ahead.

CHAPTER 6

Playing Politics: Environmentalists and the Electoral Process

Shortly after the long, lunatic farce of "Monicagate," which culminated in the acquittal of President Clinton on charges of perjury and obstruction of justice early in 1999, the 106th Congress finally returned its attention to the nation's legislative affairs. Among the first acts of the House of Representatives was to take up two statutes designed to place new restrictions on the ability of the federal government to protect the environment. One of these statutes would have precluded consideration of any legislation that would cost business an aggregate of more than $100 million per year. The other would have exempted small businesses from fines for first offenses in failure to meet environmental reporting requirements.

Just as most Republicans in Congress had ignored the clear wishes of the American people when they tried to void the 1996 plebiscite by removing the president from office, so, too, did they dismiss the wishes of the people, clearly demonstrated in poll after poll, for strong protection of the environment and public health. Congress's attempt to undercut the environmental gains achieved since the first Earth Day was part of a pattern it had followed since the Republicans gained control of Capitol Hill in 1994. Indeed, federal activism on the environment had, with only a few exceptions, slowed to a crawl since the election of Ronald Reagan to the presidency in 1980. In the arena of American politics, the environmental movement has been largely on the defensive for two decades, struggling to fend off continual attacks on the laws and institutions it helped to create with so many years of effort.

Even supposed political allies of the environmentalists have proven timid champions of environmental causes while in office. During their 1996 reelection campaign, President Clinton and Vice President Gore repeated the mantra "Medicare, education, and the environment" in virtually every speech, seeming to elevate the environment to the top of their national agenda. Yet their administration was criticized sharply for lack of courage and initiative on the environmental front. The prickly journalist Alexander Cockburn wrote in 1998 that "Clinton cannot boast of a single substantive environmental monument to his Administration."[1] Political scientist Michael Kraft said he thought that Clinton's environmental record did not deserve such harsh criticism but that "high expectations and disappointing results partially explain the phenomenon. So does the unwillingness of presidents to talk plainly with the American people about the environmental problems the nation faces and the tough choices to be made. Regrettably, the current state of U.S. politics does not encourage the public dialogue that is so clearly needed."[2]

As Walter Rosenbaum asserted, "Environmental protection, preservation and restoration ought to be explicitly recognized in public policy making . . . as issues with embedded values having a transcendent claim on national resources in competition with other values in the political process." But, he noted, despite the wishes of the American people and the efforts of the environmental movement, "the United States has been unable or unwilling to create a politics of goals and priorities in which ecology will be assigned an appropriately elevated place."[3]

Why? Why does the environment, an issue of transcendent importance, carry so little weight in the American electoral process? Why is care for our habitat, an overarching imperative that should have the potential to transform and elevate politics, so polarizing and embittering?

There is no simple answer to these questions. Economic, social, and cultural conditions all heavily influence our political system. As the Sierra Club's Michael McCloskey noted, "Environmental public policy has become a very complex enterprise." But there can be no doubt that the political process itself, the people we elect to positions of power in government and, crucially, the way we elect them are key to our public policy failures, particularly in the field of environmental protection.

The environmental movement historically has been a minor, rather ineffectual player in the electoral process. Its engagement has been tentative and diffident—it plays at politics rather than going to war. Doug Bailey, a veteran Republican political consultant who publishes the daily *Hotline* report, contended: "Environmentalists have a disdain for electoral politics. They are not good at it because they don't like it. They are pure, and purity is not a good way of achieving political goals. The leadership of the movement has to decide it wants to be a major player." And Deb Callahan, who directs the League of Conservation Voters, the political arm of the national environmental groups, said she thinks that "environmentalists are too well behaved" in politics: "We need to get more aggressive. We need to make the environment a top-tier issue." One of the few political professionals in the environmental community in the late 1990s was Philip Clapp, a former Senate staff aide who heads the National Environmental Trust. The trust was formed, he said, to help fill the political void within the movement. "The environmental community was basically sitting with the same structure of legal and scientific expertise. It had no public relations or lobbying skills, no really serious grass-roots operations except those run by volunteers who cannot run with the professionals. They were completely outgunned. This situation still has not been remedied."

By the end of the 20th century, it seemed obvious that the environmental movement could no longer count on latent public support, on scientific evidence, or on its position on the moral high ground to achieve its political goals. To address the looming threats to the planet in the years ahead, environmentalists will have to acquire the political *power* to do so—power that will put them in a position not just to ask or argue or beg but to demand that government do what is necessary.

They will need the power to hold elected officials accountable—to be able to put the fear of God into them, or, more frightening to politicians, to put the fear of losing office into them. Environmentalists will have to get seriously involved in the electoral process.

The Political Landscape

The landscape of American politics changed dramatically in the last decades of the 20th century. In response to the end of the cold war and in a continuing reaction to the cultural shocks of the 1960s and even to the progressive tide that had been running since the New Deal of the 1930s, the latent conservatism of the electorate deepened and hardened. The Republican Party was taken over by its hard-core right, and the Democratic Party retreated from its traditional progressivism. With the defection of the "safe South," the disappearance of big-city political machines, and the rapid decline of the trade union movement, the Democrats lost their long-held grasp on the majority of the electorate. The parties themselves lost control of the process to a new politics of special interests, which, as Yale University's Daniel Esty noted, is not good for the environment because it tends to produce stalemates and preserve the status quo. The political activism and idealism that boiled out of U.S. campuses in the 1960s and helped give energy to the civil rights, feminist, anti-war, and environmental movements was replaced by a self-centered quest for security and status and the accumulation of consumer goods. Spawned by venal or ideological pressures or by misinformed and misplaced populism, new political movements such as the "wise use" and property rights movements emerged to challenge the environmental agenda.

Most dismaying of all, perhaps, is that American democracy, to the extent it is defined by citizen participation in the political process, eroded perceptibly in the latter part of the century. Peter Montague, publisher of *Rachel's Environment and Health Weekly,* noted that well less than 50 percent of the electorate voted in elections, "perhaps because they see that the system offers no real choices and that voting no longer promises to make any substantial difference in their lives."[4] Although clearly unhappy with the quality of their government, Americans, enjoying relative economic comfort and in the absence of threats to their security, have been willing to tolerate elected officials who do little more than muddle through the day-to-day affairs of the nation. However, as William Ophuls and A. Stephen Boyan Jr. commented, "We Americans have taken muddling through, along with laissez faire and other prominent features of our political system, to an extreme. We have made compromise and short-term adjustment into ends instead of means, have failed to give even cursory consideration to the future consequences of present acts, and have neglected even to try to relate current policy choices to some kind of long term goal."[5]

The future consequences of present acts are, of course, what environmentalism is all about. America is unlikely to have a public policy equal to the environmental challenges of the 21st century without active participation of the electorate in the political process.

There is a variety of reasons for the disengagement of so many Americans from the process. But certainly a major reason is the highly rational belief that forces beyond their control determine the outcome of elections and the course of public policy. Chief among those forces are money, partisanship, and ideological zealotry

John Rensenbrink, a political scientist at Bowdoin College and a leader of the Maine Green Party, contended that "office holders, by and large, respond to a tune not coming from the people at large but from some few with immense resources of money and power. . . . It becomes a necessary part of successful politics to make decisions favorable, or at the very least not unfavorable, to this small power elite, and at the same time whip up the rhetoric to assure everyone and themselves that what they are doing is on behalf of the great masses. This charade is the dirty secret of both Republican and Democrat party politics."[6] The repeated contention of conservatives that government is the enemy of the people became a self-fulfilling prophecy as they took control of government.

Money and Environmental Politics

"Politics," said Will Rogers, "has got so expensive that it takes lots of money to even get beat with."[7] And that was in the 1930s.

Politics today is a lot more expensive. In the 1997–1998 fiscal year, which included an off-year national election, political action committees shelled out $230 million. Of that amount, nearly $150 million came from business organizations and $392,000 came from environmental organizations.[8] And that is only the tip of the iceberg because it does not include "soft" money contributions. In the two years leading up to the national election of 2000, which included a presidential campaign, it was estimated that candidates for office would spend some $3 billion, much of it on "uninformative" television advertisements.[9] Nor does that sum include the more than $1.25 billion spent on lobbying Congress and the executive branch by the "influence industry"—corporations, industry trade associations, and other interest groups.[10] John Stauber, editor of *PR Watch,* contended that "the corporate flacks, hacks, lobbyists and influence peddlers, the practitioners of modern PR . . . have become a kind of occupation army in our democracy."[11]

"To politicians," explained Doug Bailey, "money is the milk of politics. It is what drives political campaigns. It is a ridiculous situation, but it is true. Lobbyists give to incumbents as a way of saying 'Thank you, and we know you will be with us next year.' And if the incumbent loses, they immediately give to the winner to retire his or her campaign debt." Indeed, many, if not most, politicians spend a substantial amount of their time in office raising money for the next reelection campaign. The League of Conservation Voters has had some success in electing pro-environmental candidates and defeating anti-environmental candidates, but as Deb Callahan noted, the league has the financial resources to intervene in only a limited number of races. "It breaks your heart," she said. "Money makes an absolute difference in electoral politics." Community activist Terri Swearingen, who was deeply disillusioned when construction of a waste incinerator in East Liverpool, Ohio, went forward after Al Gore had promised to block it during his 1992 vice presidential campaign, said that in our political system, "money can make dioxin safe on corn flakes."

The environmental movement will never be able to come remotely close to matching the corporations and their auxiliaries dollar for dollar in influencing the electoral and policy-making processes. The national environmental organizations now devote only 1 percent of their budgets to politics, and those budgets are dwarfed by the financial resources available to the foes of environmental regulation. Because the system is so stacked against it, the environmental movement will not be a major factor in the electoral process unless and until the system is changed. Accordingly, campaign finance reform ought to be at the very top of the environmental movement's legislative agenda.

"It is absolutely crucial to fight for campaign finance reform," insisted California legislator Tom Hayden. "Nothing could be more important than breaking the connection between money and politics." Many leading national environmental organizations agree with that statement, but few of them are putting forth much effort to change the law. One exception is the U.S. Public Interest Research Group (U.S. PIRG), whose executive director, Gene Karpinski, said, "We put campaign finance reform at the top of our agenda." But U.S. PIRG has programs that extend beyond environmental policy. Arlie Schardt, who heads Environmental Media Services and has worked in political campaigns as well as in the environmental movement and journalism, believes that "campaign finance reform is the single biggest environmental issue in this country." But when he tried to get the national environmental groups to unite behind a campaign to reform the law, "not one of the top groups was willing to do anything about it." Why? "Inertia as much as anything," Schardt said.

In failing to support campaign finance reform actively, however, environmentalists undermine their own prospects for any significant progress. Greg Wetstone, legislative director of the Natural Resources Defense Council (NRDC) and former staff director of a House environmental subcommittee, asserted: "Campaign finance reform is vital. It is impossible for us to win, especially in committees, because we can never outbid the opposition."

Changing the campaign laws would not be a walk in the park even if the environmentalists unite with others who see the domination of politics by money as a travesty of political equality. There is, of course, a catch-22. The legislative battle needed to enact reform cannot be fought without a substantial amount of money. And those who oppose reform, today's big political spenders, are in a far better position to buy legislation. Moreover, there is a constitutional hurdle to leap. In its 1976 decision in *Buckley v. Valeo,* the United States Supreme Court ruled that unlimited expenditure of money is free speech protected by the First Amendment. The decision overturned congressional legislation that limited spending on congressional campaigns. Many legal scholars have contended that a ruling that gives more free speech to the wealthy than to other Americans is a serious misreading of what the founders intended.[12] But the decision stands.

Nevertheless, the task of reform is not impossible. By century's end, a number of states had adopted campaign finance reform laws. And the environmentalists should not be alone in the effort. Many sectors of American society who perceive themselves to be disenfranchised by the power of money in politics or who are dismayed by the inequity of a system that gives the wealthy more rights under the Constitution than the poor are seeking to change the system.

William Clark of Harvard University's John F. Kennedy School of Government contended that because campaign finance reform is "not an environmental issue per se" and because "the environmentalists would be only a tiny part of the folks working on campaign finance reform, the chances are they could do more for the environment by putting their money into saving 600 acres of forest, and so [they should] should stick to that." In this view, environmental organizations should spend their limited financial resources only on immediate environmental problems, not in trying to change the political system. The problem with that argument, however, is that the system can dictate whether and how environmental problems are solved. Environmental groups were unable to protect many areas of national forest, for example, after the 104th Congress, dominated by foes of governmental regulation, passed the so-called salvage logging rider, which gave timber companies access to previously out-of-bounds forest areas.

Money works to dilute the political influence of the environmental movement in another important way—through the tax laws. Most environmental groups rely heavily for support on tax-exempt donations. Under the current tax code, nongovernmental organizations that receive such donations are prohibited from engaging in partisan political activity. They can distribute information to help educate the electorate about issues, but the law restricts their lobbying activities and prohibits them from giving money to, or working for or against, any candidate for office. For-profit businesses have no such restrictions.

Ophuls and Boyan assert: "The gross political inequality of profit and nonprofit interests is epitomized by the favorable tax treatment accorded the former . . . the nonprofit organizations that depend very heavily on donations are severely handicapped; if they lobby, they undercut their financial support. Businesses, by contrast, can deduct any money spent for the same purpose from their taxable income and pass on the remaining expense in the form of higher prices. The public, both as consumers and as taxpayers, therefore subsidizes one side in environmental disputes."[13] As one foundation executive noted, because of the prohibitions of the tax law, "the best and brightest in the environmental movement cannot participate in politics."

This inequity should be remedied. Although the necessary changes in the tax laws were not possible in the end-of-century political morass, the environmental community, in collaboration with other communities in the voluntary sector such as community development organizations, ought to begin preparing the ground for reform immediately. In the meantime, they need to do all they can under the current statutory limitations—for example, significantly strengthening the League of Conservation Voters. The league's Deb Callahan said she is convinced that "the tax law benefits do not outweigh the damage of ceding the political field. We don't play adult politics."

Unfortunately, the mainstream environmental organizations have been made timid by the tax laws and treat them with fear and reverence, as if they are a sword of Damocles that may fall at any instant and sever the financial support provided by wealthy donors.[14] Tom Hayden, a warrior in the electoral lists for many years, said he thinks that most environmentalists just do not get it. "If government is creating the crisis, it has to be confronted politically and not just through tax-exempt organizations. There is an environmental culture that seems unable to understand what every American outcast group has learned about politics—you don't get your way by being a lobbyist dependent on the system. You get your way by threatening the system through electoral politics."

Money and Media

The biggest legislative defeat suffered by the Clinton administration was inflicted on it by a fictional couple named Harry and Louise. As the stars of television commercials invented by an advertising agency and paid for in large part by the insurance industry, Harry and Louise were the principal assassins of a health-care reform campaign that was at the top of Clinton's agenda during his first term in office.

"It is a fact of life today that the world is driven by television," observed political professional Doug Bailey. "It just is. The dominant fact in our culture is that the television set is on six hours a day. Six hours a day! If the environmentalists are going to be serious about changes they want to make or simply hold ground that has been established through previous legislation, they need to get into the battle. And the battle these days where political results and the culture are affected is on television. But the only pro-environment ads you see are by energy companies and electric utilities trying to convince you they are not bad guys."

It is another fact of life that television is a very expensive medium of communication. A few seconds of network airtime can cost many thousands of dollars, and even local airtime does not come cheap. The environmental movement cannot compete with the deep pockets of the corporations. And corporate money buys much more than commercials. Because they foot the bill, advertisers have a lot to say about the kinds of programming carried by television networks and stations and about the messages the programs convey. Because Americans spend so much of their time watching television, programming can affect citizens' political awareness, understanding, and participation. Scientist and author Sharon Beder asserted, "The media's bland diet of superficial material does not encourage political participation by the audience." Television in particular, she said, "tends to depoliticize its viewers by filling their time with mindless passive entertainment which portrays the existing system of free enterprise and consumption as generally beneficial and gives only limited time to protest groups, usually the more moderate of these."[15]

Commenting on the role of money and television in politics, Newton Minow, former chairman of the Federal Communications Commission, said: "We now have a colossal irony. Politicians sell access to something we own: our Government. Broadcasters sell access to something we own: our public airwaves."[16]

For our democratic society to work, the media must provide Americans with sufficient and accurate information to exercise their rights and duties as citizens. Increasingly, however, the mass media in this country are failing to meet that need. Television and other mass communication outlets are becoming mere vehicles for entertainment and advertising, providing little intellectual nourishment to help citizens make informed decisions about crucial environmental issues or other public policy.

Money also influences the content, including the political content, of newspapers, magazines, radio, and other mass media not only because of advertising but also because of the money corporations spend on public relations. Here again, environmentalists are outgunned—and not just by the amount of money spent. With a few exceptions, environmental organizations fail to come close to matching the professionalism and effectiveness of corporate public relations efforts.

Although the environmental movement faces a substantial economic handicap as it tries to get its political message to the public through the media, it is by no means helpless. Doug Bailey pointed out that although national environmental groups have far less money than corporations, they do have money and the potential for raising much more. They are not, he believes, spending enough of it on media campaigns, especially on television. And although money dictates much of what appears in the media, there are a good number of honest journalists covering the environment who care about getting the environmental story out and are willing to fight against the odds, including, all too often, their own editors, to do so. Although the quantity of environmental stories in the media, particularly television, declined during the 1990s, their quality steadily improved, thanks in part to the efforts of the Society of Environmental Journalists. But environmentalists will have to become much more skilled in using the media and devote more resources to conveying their message in ways that will move Americans to act politically on behalf of the environment. Certainly they can help reporters on the environmental beat by publicly holding publishers and editors accountable for neglecting or misrepresenting environmental stories and by praising media that do a good job of bringing the issues before the public. And they can learn to use the Internet effectively to mobilize the public for political action.

Alison Anderson, a sociologist who has examined the relationship between the media and environmental policy, noted: "Despite competing against economically powerful official sources, voluntary organizations can sometimes mount successful media campaigns. . . . A carefully targeted strategy aimed at mobilizing an issue to draw attention to wider concerns can be very successful in the realm of symbolic politics, despite relatively small information subsidies."[17]

Parties, Partisanship, and Ideology

Once upon a time, and not so long ago, the environment was a unifying issue in American politics. In the years following the first Earth Day, the discourse on environmental policy was conducted to a surprising degree in a context of cooperation, consensus, and bipartisanship. Even President Nixon, an immoderately partisan Republican, found it politically expedient to be a strong supporter of efforts to protect the environment. In the mid-1970s, legislators such as Edmund Muskie, a Democrat from Maine, and Robert Stafford, a Republican from Vermont, worked together smoothly and effectively on the Senate Environment and Public Works Committee to draft, enact, and strengthen the great environmental statutes. Some of the strongest environmental advocates in Congress, including Senators John Sherman Cooper of Kentucky, John Heinz of Pennsylvania, Jacob Javits of New York, Charles "Mac" Mathias of Maryland, and many more, were Republicans.

Bipartisanship regarding environmental issues—and virtually all other issues—faded, however, with the rise of the Republican right and as the party of Theodore Roosevelt became the party of Ronald Reagan. The anti-government, laissez-faire ideology of the right was directed with particular virulence at environmentalism, which was treated as the tooth-and-claw enemy of the supposedly God-given right of industry and property holders to exploit, pollute, and generally operate free of legal restraint, regardless of the consequences for humans and their habitat. Reagan's appointees to key environmental positions, especially Secretary of the Interior James Watt and Environmental Protection Agency administrator Anne Gorsuch, launched a fierce campaign to get government "off the backs" of Americans by easing or abandoning enforcement of environmental laws, trying to weaken the laws, and turning over public property to private interests.

Because of effective opposition by the national environmental groups, which alerted and aroused the American people, the Reagan administration, in Philip Clapp's words, "went dark" on the environment in less than four years, getting rid of Watt and Gorsuch and ending its overt assault on the laws. However, Clapp noted, industry quickly stepped in and "threw a ton of money" at thwarting environmental laws and agencies with grass-roots organizing efforts and phony citizens' groups and, in an "unholy alliance" with political consultants, accelerated its efforts to control the electoral process.

A decade later, with the right wing firmly in control of the party, the Republican-dominated 104th Congress, with Newt Gingrich's "Contract with America" as its manifesto, again mounted a full-scale assault on environmental laws and institutions. Intended as a major weapon in that assault was the proposed Job Creation and Wage Enhancement Act, which would require what Michael Kraft described as "stifling" cost–benefit analysis of all federal regulations. The title of the bill, which was supported by the same Republicans who opposed any increase in the minimum wage, backed anti-union legislation, and did nothing to oppose corporate downsizing and exportation of jobs, was of course cynical and deceitful. The legislation was aimed at paralyzing the federal regulatory effort, not by any stretch of the imagination at helping workers.

Again, though, not all Republicans were anti-environmental extremists. The Republican moderates left in Congress, led by Congressman Sherwood Boehlert of New York and Senator John Chafee of Rhode Island, did what they could to help stem the tide. Several Republican governors, such as George Pataki of New York and Christine Whitman of New Jersey, distanced themselves from their party's efforts to roll back thirty years of environmentalism. Early in 1999, Republican governor Tom Ridge of Pennsylvania wrote in an op-ed article for the *New York Times:* "Too many Republicans have allowed the party's bedrock principle of limited government to excuse the absence of an aggressive, creative environmental strategy—resulting in a widely held belief that Republicans are more interested in protecting big business than in preserving parks, air, rivers and forests."[18] But Congress and the party were under the control of the right-wing ideologues, to whom restrictions on economic growth or corporate freedom of action for environmental reasons is anathema. Key environmental positions were almost uniformly in the hands of anti-environmentalists, including Tom DeLay, Dan Burton, Don Young, David McIntosh, Frank Murkowski, Trent Lott, Jesse Helms, and their ilk. In a chilling echo of the McCarthy-era witch hunts against supposed communists, Republican congressman Young of Alaska, chairman of the House Committee on Resources, demanded to know whether any federal employees were members of or had any contact with environmental organizations.[19] Because ultraconservatives dominated the political primary apparatus, moderates faced the alternative of toeing the hard-right line or possibly losing their seats in the next election cycle. Russell Peterson, former president of the National Audubon Society and a former Republican governor of Delaware, was so disgusted with the anti-environmentalism of the Republicans that he switched his affiliation to the Democratic Party.

Republican conservatives also became increasingly influential in statehouses and on school boards and local county councils as the Christian right sent candidates into the electoral lists. The "conservative" appellation was often a misnomer as the right pursued an agenda that focused on radical goals, including rolling back the environmental laws. By their ability to mobilize hard-core conservative voters, the right wing of the party was able to defeat moderate Republican candidates in primary elections or cow them into toeing the right-wing line. In state legislatures, the conservatives were able to block or roll back environmental legislation. At the local level, environmental education was curtailed and textbooks barred by right-wing-dominated school boards. Conservative appointments to the bench at every level of government diluted enforcement of environmental laws.

The scorecard kept by the League of Conservation Voters to track environmental votes showed that the great majority of congressional Republicans consistently cast anti-environmental votes, averaging less than 20 out of a possible score of 100. Of course, as former senator Gaylord Nelson, now a counselor for The Wilderness Society, noted: "No one in Congress runs as an anti-environmentalist. They say, 'But I am not a crazy environmentalist who will take your job away.'" The minority Democrats in Congress, meanwhile, scored on average more than 80 percent on the League of Conservation Voters' scorecard. But the best the Democrats have been able to do in recent years was turn back some of the excesses of the Republican right—and then only because the environmental organizations were able to get their act together long enough to mobilize public opinion to send a clear message to members of Congress.

In addition to being in the minority, the congressional Democrats lost much of their environmental leadership in the 1980s and 1990s. Gone were the likes of Ed Muskie, Frank Church, Morris Udall, Phil Hart, Tim Wirth, Phil Burton, Jim Florio, and other environmental activists on Capitol Hill. Only a small, overmatched rear guard, including Congressmen Henry Waxman and George Miller, Senator Paul Wellstone, and a few others, remained. Although they were not ideologically anti-environmental, members of the Democratic minority in Congress were ineffectual defenders of environmental laws and institutions.

Nevertheless, the Democrats were finding that the environment was an increasingly effective issue for them in election campaigns because of the excesses of the Republicans. As Greg Wetstone observed, "The environment is becoming a core Democratic issue." It was a bit surprising, therefore, that the Democratic Party and its candidates did not grasp the environmental issues more firmly and aggressively.

The sharp polarization of environmental policy along party lines is a problem for the environmental community. Environmentalists have always insisted that their cause is not a partisan issue, that a healthy, attractive environment has nothing to do with traditional liberal–conservative economic and social alignments. Their political organs usually leap at the chance to endorse a Republican with a decent environmental record in order to prove their nonpartisanship. But facts are hard to avoid. The Republican Party has become the anti-environmental party in recent years. Chuck McGrady, a Republican who became president of the Sierra Club in the late 1990s, warned that his party "will face the continued wrath of dissatisfied voters" unless it aligns itself with the majority of voters who identify themselves as environmentalists. He admonished the party's representatives in Congress to "stop subverting the democratic process" by attaching to bills anti-environmental riders that otherwise have no chance of passing.[20]

If environmentalists are to hold their own in the political arena, they will have to drop the fiction that environmental policy is bipartisan until it becomes so in reality once again. They need to get more deeply into the down-and-dirty political work of punishing their enemies and rewarding their friends where it counts most profoundly: in the electoral process. To do so, they will need more political clout, better strategies, and strong allies, and they will need to put far more effort into politics than they currently do.

Getting into the Game

The environment is something almost everybody talks about and almost nobody does anything about in the voting booth. Public opinion polls consistently indicate that most Americans, more than three-quarters of them, support the environmental enterprise. But green issues rarely are high on the list of reasons a voter considers when casting a ballot. There are a number of explanations for this, including the role of money and media discussed earlier in this chapter. Another explanation, however, is that environmentalists have failed to tap the latent political support suggested by the polls.

"We have power; we are just not exercising that power," said Deb Callahan of the League of Conservation Voters. "If we could deliver on that power, the politicians would be beating at our front door. When we show muscle, it works, but mostly we [are] writing checks to nice candidates."

To exercise power, the environmental community will have to do something it has largely neglected in the past—*organize* for power. This will require environmental groups to raise much more money and then concentrate that money on the electoral process. It will mean building an army of grass-roots political organizers, both volunteers and paid staff, who will operate in every state and in thousands of communities, going door to door during election campaigns, educating voters and getting them to the polling place. "We absolutely need more politics, more political organizing," said Lois Gibbs, director of the Center for Health, Environment, and Justice. "That's how the Christian Coalition got where it is. They put their people on school boards, on city councils, and in statehouses. We have to see how they have done it and do the same."

Only a handful of environmentalists have the political skills to get into the pit with professional political operatives. A critical but neglected task of the environmental organizations is to train a cadre of organizers, media specialists, and political fund-raisers. As author David Korten noted, one of the most important roles of nongovernmental organizations in a democracy is to "provide a training ground for democratic citizenship, develop the political skills of their members, recruit new political leaders, stimulate political participation, and educate the broader public on a wide variety of public interest issues."[21]

A number of the environmentalists, political experts, and scholars interviewed for this book said that the poor political record of the environmental community can be explained by the poor job it does in demonstrating how its agenda links up with the real problems and needs of the American people. To make that case, national organizations will have to engage much more intensely than they now do in the politics of local communities—in the "politics of place," to use Daniel Kemmis's term. Although the polls may show that people are concerned about the environment, the environment as an abstraction fails to motivate electoral decisions. But people do care deeply about concrete environmental issues where they live. They may be unmoved by the national problem of acid rain but will become aroused by dead trout in their local streams and dying trees in nearby forests. Residents of inner cities may not care about national ambient air quality standards, but they will fight—and, if organized, vote—to stop the pollution that is giving their children asthma. Environmentalists will have to address environmental problems at the global and national levels, but to do so they must adhere to the dictum of the late Tip O'Neill that "all politics is local."

One good place to look is the Rocky Mountain West, an area that is so dissatisfied with federal policy that thoughtful people such as Kemmis are pressing for regional autonomy. If all the talk about the "New West" and the people moving there for its beauty, recreational opportunities, open space, and fresh ecology has any basis, the region ought to be fertile ground for environmental politics. Instead, political power in the region remains largely in the hands of the traditional extractive industries, which make sure that conservative politicians are sent to Washington, D.C., and to statehouses, politicians who will not let environmental laws and institutions interfere with industry's ability to exploit the area's natural wealth. They have been successful in doing so at least in part because the national environmental movement has done little to understand the needs and views of that section of the country or to help regional and local activists organize the latent political strength of New West voters.

At the end of the 20th century, the environmental community lacked—by orders of magnitude—the financial resources, the personnel, and the skills and training to engage in the broad political effort described here. But we are talking about the 21st century. The great environmental challenges of the new century will not be met unless environmentalists acquire the political strength to do so—to turn broad cultural and social acceptance into political power. They will have to acquire the wherewithal to become a political force.

Even, however, if the environmental community acquires the tools to become an effective force in the electoral process, it will never have sufficient power to turn the tide on its own. It will have to find allies, to work in coalitions for common political goals.

Joining Forces

Environmentalists have rarely been good at forming and working in coalitions, even within their own ranks. The national organizations can come together in emergencies, such as the need to repel the attack on environmental laws, or in a drive toward a goal of exceptional importance, such as the campaign that produced the Alaska National Interest Lands Conservation Act, but they are usually too competitive or preoccupied with different goals and strategies to work in close cooperation over an extended period. Although the "Green Group" of twenty-plus environmental and related organizations meets to discuss issues and tactics, it does not represent an organized core of the environmental movement—certainly not when it comes to electoral politics. There have been infrequent efforts to mobilize state and local grass-roots organizations for ad hoc political campaigns around a piece of legislation here or a piece of real estate there. Yet the thousands of local organizations such as those that work with Lois Gibbs's Center for Health, Environment and Justice, many of which have engaged in fierce local political firefights and acquired substantial political skills, could be the muscle for a powerful political movement.

One cannot say that the environmentalists are failures at forming broader political coalitions with other sectors of American society because they have almost never really tried. Kalee Kreider of Greenpeace believes that environmentalists "don't do coalitions well because we focus only on our own expertise. It limits us from thinking outside the box." On the other hand, John Passacantando of Ozone Action does not think that getting together with those in other sectors to work on a broad social agenda is the job of environmentalists such as himself. "There is not much I can bring to it," he said.

Stephen Viederman of the Jesse Smith Noyes Foundation is convinced that environmentalists cannot be successful at coalition building *because* they do not concern themselves with broader social issues. Many environmentalists and environmental organizations do not consider protection of the environment to be a liberal or progressive movement and do not want to be labeled as such. NRDC's Greg Wetstone noted that environmentalists had joined with unions to beat back attempts to cut worker health and safety protections and had also tried to focus on environmental justice issues. He added, however, that these issues cause tension among many in the environmental movement "who will tell you they need to reach out to a broader political spectrum and not be pigeonholed as part of the liberal community." David Hawkins, a senior attorney for NRDC and one of its founders some thirty years ago, pointed out that the general membership of most national environmental organizations will respond with greater support for narrower, more clearly articulated goals than for a broad social agenda.

The leaders of the mainstream environmental organizations are, as a rule, reluctant political risk takers, unwilling to entangle their agenda with broader social and economic issues. The largely white, middle-class leadership and cadre of the groups often seem to bend over backward to prove that they have no desire to change the status quo except when it comes to their own specific pollution, land, and resource issues. They want to be seen as sociologist Denton Morrison described them: "the safest movement in town."[22] They are concerned, David Hawkins noted, with issues of distributive justice toward future generations but neglect distributive justice for the current generation. They shun the labels "liberal" and "progressive" as if they were a deadly virus. Indeed, some liberal Americans are critical of environmentalism and environmentalists for supposedly diverting resources and public attention from such issues as social justice, education, and other items on the "liberal" agenda. And yet, Hawkins said, many environmentalists of his generation, who joined the movement at the time of the first Earth Day, did so "to pursue an agenda of aggressive reform, including racial, economic, and gender justice and worker protection and rights."[23]

Some radical environmentalists—those willing to shake up the existing order—such as Earth First! do not have any kind of people-centered social and economic agenda. Others, the social ecologists, for example, who are affiliated with Green parties around the country, do have a vision of a more just and democratic society but remain a marginal part of the movement. More than any other sector of the movement, they actively seek political office as a means of achieving their goals. Perhaps the closest thing to coalition politics practiced by environmentalists are the efforts of the grass-roots organizations that work with Lois Gibbs's center. As Gibbs frequently says, all local battles over the environment are political. Such battles are winnable only through the cooperation of a wide spectrum of the community in which they are fought.

Although it is said that the United States is a conservative country, the four decades that preceded the first Earth Day were dominated by progressive politics supported by a relatively unified electorate. Those decades witnessed the great social reforms of the century, including the emergence of Social Security, strong labor laws, landmark civil rights statutes, long strides in providing decent housing and health care for Americans, and a list of other programs designed to ensure that all citizens share in the nation's affluence and security. The landmark environmental statutes were enacted by politicians whose careers began and ripened in those years.

That burst of governmental activism, which transformed the nation into a more just and secure society, was enabled by a political coalition that included industrial workers and their unions, big-city politicians representing immigrant and minority communities, agrarian populists, professionals and academics, and an emerging, newly affluent middle class. It was an unstable coalition, with frequent defections and shifting alignments depending on the issues before it. But it was strong and durable.

However, this progressive coalition crumbled under a variety of blows, including the "white flight" to the suburbs following the great African American migration from the South, the split among Americans over the Vietnam War, the rise of feminism, the debilitating drug epidemic, the decline of rural America, the new sexual freedom and other lifestyle changes that challenged traditional moral values, the decline of the trade union movement, the rise of the global economy and the disappearance of many basic industrial jobs, the intense polarization of abortion politics, and mounting antagonisms among former allies such as Jews and African Americans over such programs as affirmative action. The decline of the progressive impulse in American life was accelerated by the political aggressiveness of Christian fundamentalism; the weakness and disunity of political parties, particularly the Democratic Party; and the growing disparity of wealth among Americans, which was both a cause and an effect of the decline. And those are only some of the reasons for the end of the old coalition.

Whatever the reasons, the forces and alliances that created the political context for the emergence of contemporary environmentalism have largely evaporated. Public support may be there, but the infrastructure of political progress is not—environmentalists have not created a mechanism for using that support to put pro-environmental candidates into public office. If environmentalists require political power to achieve their goals, as I believe they do, they will have to enlist allies, to find or create a new coalition.

With whom can they form alliances? The corporations? Perhaps they can ally with industries and businesses that have a stake in a greener economy, but certainly not with those that now oppose efforts to reduce consumption, resource depletion, pollution, and erosion of life-support systems. Environmentalists need to find common ground with others in American society who are seeking to change the political, economic, and social status quo, those who are not secure and comfortable but who suffer from inequity or see dangers for our democratic society as well as dangers for our physical environment down the road.

Although the burden of a half century of political history almost certainly precludes the resurrection of the old progressive coalition, some of its elements can be enlisted in a new alliance, including the industrial unions and some of the newer service industry and teachers' and farmworkers' unions. The environmental justice movement is the civil rights impulse in another incarnation and embodies high political energy. Scholars and professionals who understand the seriousness of the environmental threats we face, particularly health professionals and scientists who are members of organizations such as Physicians for Social Responsibility and the Union of Concerned Scientists, can be active and effective members of a political alliance. The thousands of grass-roots environmental groups around the country represent a potential army of frontline troops who have already acquired some political skills. So, too, do the hundreds of thousands of people in the community-based development movement discussed in chapter 4. The intensifying attention to ecological theology suggests that religious leaders and their followers can be enlisted to lend ethical and spiritual support to a new progressive coalition. The swelling ranks of retired Americans who are concerned with lifestyle issues, including the quality of their physical environment, are likely recruits. Add all these to the environmentalists' base in the white suburbs and there is potential for formidable electoral power.

Given the present fragmentation of interest politics in the United States, it will not be easy to bring these elements together. The alliance will have to be built brick by brick, and the environmental movement, with its broad latent popular support, will have to shoulder much of the construction work. And to draw all the diverse elements into alliance, environmental groups will have to expand their agendas to accommodate such traditionally progressive goals as distributive justice. William Shutkin of Boston's Alternatives for Community & Environment asserted that "we need more environmentalists who, as a matter of default, are talking about democracy, about economic development, as a central aspect of environmental protection." Philip Clapp said, "We have an alternative of creating a progressive agenda as opposed to sitting all by ourselves and trying to put a cap on every [source of pollution]."

If the mainstream environmental groups do not want to be pinned down by the "progressive" label, they can describe the alliance with any term they like; they could call it the Pro-Life Coalition, a term certainly more descriptive of environmentalism than of a movement that countenances murder in the name of saving fetuses. As Karl Grossman, a teacher and author, observed, "The environmental crisis is about the right to life." But progress means forward movement, and if environmentalists are to move toward their goals in the 21st century, they will require a progressive politics with which to do so, whatever it is called.

The Greens

A discussion of environmental politics in the United States must at least touch on the activities and fortunes of the Green parties. Because this country rejects proportional representation in favor of a winner-take-all political system, third parties have had little chance of achieving significant power within government. And because the Greens in the United States have been so divided and contentious and command so little money and media coverage, they have not so far been taken seriously as a factor in American politics. When consumer advocate and activist Ralph Nader ran for the presidency on the Green Party ticket in 1996, he won a negligible number of votes.

But the Greens have stuck to it at the state and local levels and have been recording some modest victories. In 1998, 125 candidates from Green parties ran for office and 18 were elected, most of them to city councils. Although he lost, former Democratic congressman Dan Hamburg won more than 100,000 votes as the Green Party candidate in California's gubernatorial election, and the party's candidate for the office of lieutenant governor, Sara Amir, won nearly 250,000 votes.[24] Even without significant victories, the Greens are starting to be noticed by the political establishment, if for no more than their role as potential spoilers. In 1997, Green Party candidate Carol Miller received 17 percent of the vote in a special election to fill the seat of Democratic representative Bill Richardson in northern New Mexico. Her votes came mostly at the expense of the Democratic candidate and handed the seat to the conservative Republican candidate, Bill Redmond, in a district that traditionally returned Democrats to Congress.[25]

Beyond running candidates, however, the Green parties offer an alternative vision of what politics can be to a public that is manifestly dispirited by the me-too, money-driven politics of the two major parties. Ralph Nader said it in his usual understated way when accepting the Green Party presidential nomination. The broad-ranging program of the Greens, with its proposals for social justice, civil liberties, and civil rights as well as ecological issues, cannot be compared, he stated, with "the flaccid, insipid, empty, cowardly platforms of the Democratic and Republican Tweedle-Dum and Tweedle-Dee parties."[26] John Rensenbrink, the political scientist from Bowdoin College who ran for a Senate seat on the Maine Green Party ticket in 1996, noted: "New parties have always come forward when the established parties become entrenched, greedy, unhearing. We need to convince the public that politics is relevant. We need to change the culture, to confront the media oligarchy that controls the way people think about these matters." And Karl Grossman pointed out that on occasion in the nation's history, alternative political movements have changed national politics without winning control of the political apparatus. "Think of Eugene Debs, think of labor at the turn of the last century!" he said.

American Green politics is based on ten "key values" adopted at a meeting of the Greens in St. Paul, Minnesota, in 1986. They are "ecological wisdom, grassroots democracy, personal and social responsibility, nonviolence, decentralization, community-based economics, postpatriarchal values, respect for diversity, global responsibility and a future focus."[27] The late Willis Harman, a social scientist and futurist, saw the Green parties not only as a new kind of politics but also as a portent of a new way of thinking about the interrelationships among humans and between human society and the natural world. It is a politics, he wrote, based on "the interconnections among principles of ecological wisdom, sustainable peace, an economy with a future, and a participatory democracy with power channeled directly from the grassroots level."[28]

John Rensenbrink said that he thinks it possible that Green parties will move closer to the mainstream of American politics in the future. The Greens are pressing to shift the system to proportional representation. Some Greens think that their party will eventually replace the Democratic Party. Others think they should forget about winning or losing and just try to convince Americans of their goals. The very process of running for office, Rensenbrink said, brings respect to the candidates and a hearing for their platform.

It is highly unlikely, however, that the current political structure in the United States will change to accommodate the Greens or other third parties. The victories of Green Party candidates probably will remain limited to scattered local elections. The national environmental groups keep their distance from the Greens and are no doubt wise to do so. For the foreseeable future, the only practicable path to electoral success will be through the established parties. John Flicker, president of the National Audubon Society, observed: "We are mainstream. We have a majority of voters who share our values. Separating ourselves out would marginalize us. We should look like a majority, act like a majority, and organize our constituencies like a majority."

Unfortunately for the environmentalists, however, the majority that shares their values usually does not elect politicians who will implement those values. One thing the mainstream groups might do to turn their political energy from latent to actual would be to learn from the message of the Greens and emulate the passion with which the Greens bring that message to the voters.

The end of the 20th century was not the best of political times in the United States. Although we have enjoyed a long period of economic prosperity, have not had to fight a major war, and are unchallenged as the world's mightiest nation, as a people we are experiencing great unease over the state of our politics and our government. The lack of participation in the electoral process suggests a widespread cynicism and resignation among the electorate about the future of the democratic process. In part, that can be traced to the frailties and ineptitude of our politicians, the polarizing effects of mindless ideological partisanship, the corrupting influence of money on our politics, and manipulation of the political process by those who already have money and power. But it is also likely that the absence of a crisis, the lack of a cause, has left a vacuum at the center of our political life. For almost the first time in the century, there was no unifying goal in our politics, no call to make the world safe for democracy, no Great Depression to end, no fascism or communism to fight.

Environmental politics could provide that center. A political platform that combines social and economic justice for all people with care for the planet could send a powerful, persuasive, and unifying message to the electorate.

As scholar Lynton Keith Caldwell observed: "To argue that politics will never change is to imply that societies cannot or will not learn—that the possibility of better politics is an illusion. . . . The environmental movement and the developments that brought it into existence provide persuasive reasons for believing in the possibility of a new politics that is more than rhetorical promises."[29]

In his study of American politics titled *The Deadlock of Democracy,* political scientist James MacGregor Burns wrote that although people used to say that the cure for democracy was more democracy, "hard experience has shown this cliché to be a dangerous half-truth. The cure for democracy is leadership—responsible, committed, effective, and exuberant leadership."[30]

Burns was talking about leadership from politicians, elected officials, occupants of the White House and Congress—leadership from the political parties. But that was nearly four decades ago, and such leadership failed to emerge in the intervening years. It would seem, then, that leadership to cure our ailing democracy will have to come from elsewhere—from its citizens and their institutions. And the environmental movement is well positioned to provide such leadership in alliance with other sectors of civil society. But whether the leadership of the environmental movement is yet willing to think that broadly about its role, much less to take up the challenge, is an open question.

Chapter 7
Taming the Genie: Science, Technology, and Environmentalism

In 1998, eminent Harvard biologist Edward O. Wilson published a book titled *Consilience* that was somewhat controversial within scientific circles. "The central idea of the consilience principle," Wilson explained, "is that all tangible phenomena, from the birth of the stars to the workings of social institutions, are based on material processes that are ultimately reducible, however long and tortuous the sequences, to the laws of physics." He also noted, "As the century closes, the focus of the natural sciences has begun to shift away from the search for new fundamental laws and toward a new kind of synthesis—'holism' if you prefer—in order to understand complex systems."[1]

Several members of the scientific community publicly assailed Wilson's view as heresy or, even worse, as politically incorrect. Nonscientists attacked his assertion that spiritual values would have to stand the test of empirical inquiry and that the progress of science would eventually lead to the secularization of religion itself.[2] But the notion of consilience uncannily echoes a flash of transcendental insight that John

Muir, the spiritual father of modern environmentalism, recorded in his journal more than a century ago: "When we try to pick out anything by itself, we find that it is bound fast by a thousand invisible cords that cannot be broken to everything in the universe."[3]

Wilson's perspective is materialistic rather than spiritual. Yet it points to a significant evolution in both the ethical and intellectual orientation of science. In the latter part of the 20th century, science and technology or, at least, a growing number of their practitioners, were moving from a reductionist to a holistic view of the world—and of their place in and duties toward the world. It had been a long, slow journey.

Science, Technology, and the Environment

Francis Bacon, a creator of modern science, wrote in 1620, "The real and legitimate goal of the sciences, is the endowment of human life with new inventions and riches."[4] In his essay "The New Atlantis" written three years later, he envisioned a happy, prosperous nation governed by scientist-rulers whose labors were devoted to the control of the nature. The chief of what could be described as the country's central research and development institute, called "Salomon's House," explained to foreign visitors, "The end of our foundation is the knowledge of causes and secret motions of things; and the enlarging of the bounds of human empire, to the effecting of all things possible."[5] What Bacon foresaw, wrote historian Theodore Roszak, was the possibility that "given sufficient technological power over nature, the hope of democratic abundance might not be unrealistic. Bacon's New Atlantis holds its place as the first scientific Utopia, a bold prediction of good things to come based on the assumption that the unlimited proliferation of material goods is within the realm of the possible. That vision has hovered in the background of the entire industrial process as its one justification for the privation, harsh discipline, wrenching dislocation, grime and soot that this great adventure has cost."[6]

Science and technology have indeed produced a cornucopia of material abundance for a substantial portion of the human race, matching even Bacon's expansive vision. The modern world was essentially created by science and its functional offspring, technology. Our civilization would not have been built and could not survive without them.

Just think of the anxiety and uproar created at the turn of the millennium by a single technological problem, the Y2K computer glitch, only one small thread in the intricately woven fabric of our techno-industrial society. Science and technology have provided their possessors not only with material abundance but also with power—political and military power and the ability to impose their will on those who lack equivalent scientific and technical capacity. And they have presented humanity with broad dominion over nature.

It is little wonder, then, that over the past two centuries many people afforded science the same reverence reserved for deities in earlier times, regarding it as "the unique bearer of the True and therefore the Good."[7] Or, as Willis Harman noted: "We in modern society give tremendous prestige and power to our official, publicly validated knowledge system, namely science. It is unique in this position; none of the coexisting knowledge systems—not any system of philosophy or theology, not philosophy or theology as a whole—is in a comparable position. Thus it is critically important—to an unparalleled degree—that our science is adequate."[8]

In recent decades, science has begun to slip from its lofty pedestal as it has become apparent that it is not adequate either to meet all the needs of humanity or to protect us from the dangers that science and technology themselves create. That realization had been building for many years, but it became inescapable after the horrors of Hiroshima and Nagasaki. In the United States in particular, people have marveled at and reveled in the material wealth, personal freedom, improved health, and prolonged life span provided by science and technology. Increasingly, however, reverence is turning to dismay as we discover that the genie of science and technology is threatening the biological, chemical, and physical systems that support life and evolution.

As landscape architect Robert L. Thayer Jr. observed, "Americans, in increasing numbers and intensities, feel guilty about what technological development has done to the landscape, to 'nature,' and to the earth."[9] Along with guilt came a diminished if not shattered faith in the ability of science and technology to address all our problems and lead us to that Utopia promised by Francis Bacon.

This erosion of faith in the outcome of science and technology is accompanied by diminished confidence in the wisdom and altruism of scientists and engineers and by a questioning of the scientific method itself. In the traditional view, wrote historian Stephen F. Mason, "the scientific method relies upon rational arguments rather than emotional appeals, and it suggests that empirical evidence should decide between rival ideologies."[10] In this view, the scientist, free of preconceived values, seeks the truth and follows it wherever it leads. It is assumed that whatever the outcome of the search, it will benefit human welfare.

The reality, as Lester Milbrath asserted, is that scientists who believe their work is "value free" are laboring under a "delusion."[11] Every scientific discipline is laden with its own set of values and methods. All scientists bring to their tasks predispositions arising from their specific background and training; from financial and economic considerations, including funding sources; from peer expectations, personal ambitions, and institutional settings; and (dare it be said?) from ethical and moral beliefs—or the lack of them. It would be safe to say that scientists can be no more free of their own biases, preferences, and incentives than can journalists be fully objective.

In the face of mounting empirical evidence about the unintended destructive effects of science and technology on the physical world and its living inhabitants, a growing number of scientists and scientific institutions have concluded that simple devotion to their discipline is an inadequate value and are bringing ecological values to their endeavors.

Modern environmentalism itself, of course, is founded on a base of science. It was scientists such as Rachel Carson and Barry Commoner who raised the early alarms about the dangers to the natural world and human well-being caused by human industrial and technological activity. Scientists in government, in universities, and in research institutes alert us to the sources of environmental degradation and propose countermeasures. Many scientists today, including E. O. Wilson, Paul and Anne Ehrlich, Peter Raven, George Woodwell, and hundreds of others, are eloquent advocates for the environment. The Union of Concerned Scientists, formed to alert the world to the dangers of nuclear disaster, has long since expanded its agenda to deal with threats to the global environment. Physicians for Social Responsibility has sought to bring medical science to bear on the environmental health problems caused by the by-products of science and technology. In 1993, a group of some 1,600 scientists, including many Nobel laureates, issued a "World Scientists' Warning to Humanity"[12] cautioning that "human beings and the natural world are on a collision course" and adding that "a new ethic is required—a new attitude toward discharging our responsibility for caring for ourselves and the earth." The warning pointedly states that the help of the world community of scientists is required to meet the threat.

In 1997, naturalist George Schaller spoke for a widening circle within the scientific community when, in a speech presented to scientists and politicians at the Library of Congress, he said: "To understand nature is not enough: we all have a moral obligation to help protect what we study. . . . Many of us use science to enhance our role as ecological missionaries. The goal is to balance science and advocacy."[13]

That same year, Jane Lubchenco, then president of the American Association for the Advancement of Science, proposed a new "social contract" for the scientific community to address "the urgent and unprecedented environmental and social changes" caused by the effects of human activity on the planet. She elaborated:

> This Contract would more adequately address the problems of the coming century than does our current scientific enterprise. The Contract should be predicated upon the assumptions that scientists will (i) address the most urgent needs of society, in proportion to their importance; (ii) communicate their knowledge and understanding widely in order to inform decisions of individuals and institutions; and (iii) exercise good judgment, wisdom and humility. The Contract should recognize the extent of human domination of the planet. It should express a commitment to harness the full power of the scientific enterprise in discovering new knowledge, in communicating existing and new understanding to the public and to policy-makers, and in helping society move toward a more sustainable biosphere.[14]

Lubchenco also stated: "As we begin to appreciate the intimate fashion in which humans depend on the ecological systems of the planet, it is becoming increasingly obvious that numerous issues we have previously thought of as independent of the environment are intimately connected to it. Human health, the economy, social justice, and national security all have important environmental aspects whose magnitude is not generally appreciated."[15]

In other words, "when we try to pick out anything by itself, we find that it is bound fast by a thousand invisible cords that cannot be broken to everything in the universe." Here is transcendental holism or, as E. O. Wilson sees it, secular consilience emerging in the mainstream of American science.

If such a social contract were to be adopted by the full scientific community, not only in the United States but throughout the world, it would represent a giant stride toward taming the scientific and technological genie that has caused so much damage to the world while it was creating so many beneficent wonders. Unfortunately, much of the nation's and the world's scientific capacity remains in the hands—and pockets—of those who have no interest in setting that kind of course.

Science for Sale

What was billed as the "the largest study" ever on the effects of occupational exposure to polychlorinated biphenyls (PCBs) was published early in 1999 in the *Journal of Occupational and Environmental Medicine*. The study found no evidence of "significant" links to cancer deaths among workers exposed to PCBs on the job. The study was funded by the General Electric Company (GE), which had been fighting for years to avoid cleaning up PCBs it had dumped into the Hudson River in New York, the Housatonic River in Massachusetts, and other sites, including leakages into the ground under homes in Pittsfield, Massachusetts. The cleanup would cost, according to news accounts, hundreds of millions of dollars.[16] The company had already spent what must have been millions of dollars fighting regulatory compliance with pollution laws.

Two scientists belonging to a private institute in Washington, D.C., conducted the study for GE. The two had, in the past, been employed by the Environmental Protection Agency. When I covered that agency as a reporter, I never had occasion to doubt their competence or integrity. But journalists are always skeptical of scientific reports paid for by a commercial interest that has a financial stake in the outcome of the study. For that matter, scientific evidence produced by a non-profit organization such as Greenpeace, with its own agenda to advance, however worthy, should also be taken with a grain of salt. Even reasonably honest scientists have to consider on which side their bread is buttered, especially if they choose to take the money of a big, rich, powerful institution such as GE, which has a reputation for pushing others around rather than itself being pushed around. In many cases, scientists hired to produce specific results do not or cannot look at the broader implications of their work. As political scientist Lynton Caldwell observed, "The professional and economic interests of scientists and engineers employed in industrial and agricultural production may be threatened by the standards sought by environmentalists and consumer advocates."[17] For example, Stephen Viederman recounted what happened when he asked scientists of the Monsanto Company who worked on synthesizing hormones to increase the milk production of cows whether they had considered the health and social implications of their research. They replied that "they had no responsibility to consider these so-called by-products of their work."

Science has been invoked by corporations, and by anti-environmental ideologues, to question the seriousness and even the existence of environmental threats such as global warming. It is used to suggest that hazardous substances are not harmful, to press ahead with new, potentially dangerous technologies. It has been used to cover up dangers to human health caused by commercial activity—for example, the "medical" evidence presented by tobacco companies to argue that smoking is not harmful. It is used to justify rapid depletion of nonrenewable resources. It is used to avoid compliance with environmental regulations and to oppose enactment of new environmental laws. It has been used to suggest that our environmental problems can be easily addressed with cheap technological fixes.

Often, these activities are justified under the sobriquet of sound science. But the term is often used, as consumer advocate Ralph Nader described it, as "basically a semantic camouflage for the corruption of science by money at the behest of corporate policies that are increasingly lawless and increasingly bold."[18] Right-wing ideologues, such as House Majority Whip Tom DeLay of Texas, who ran an exterminating business before becoming a politician, parrot the message of polluting industries by describing critical problems such as the hole in the ozone layer and global warming as "junk science."[19] But Nader contends that the real junk science is the science paid for by commercial interests: "Just like bought-and-paid-for politicians and bought-and-paid-for media, bought-and-paid-for scientists are an instrument of corporate power. They are integral instruments in deferring the consequences of reality and warding off the mobilization of the citizenry and the application of regulatory health and safety laws."[20]

A good portion of this kind of science has been used in what Paul Ehrlich has described as the "brownlash" by economic interests that oppose green policies by trying to convince the public and policy makers that "environmental problems are trivial and/or environmental battles have been won. . . . One should not make the mistake of thinking this is a scholarly controversy. The brownlash is simply dirty politics with the agenda of promoting looting and polluting for profit. The short-term gains of the few will be paid for by long-term losses for everyone."[21]

This is certainly not to suggest that all scientists who take money from commercial interests are prostituting their scientific integrity. It is possible, for example, that at least some of the small cadre of scientists who have recently made a career of trying to debunk the scientific consensus on global warming do so because they truly disbelieve the evidence, not because they are receiving money from the fossil fuel industry. Undoubtedly, many of the scientists who work for corporations do so to make a contribution to human welfare, and often they do just that. But it is inescapable that, as Lester Milbrath observed, in today's world "the control of science and the power it creates will go to those social entities that control funding for its development. In a market society, most control will flow to large corporations." In the face of such power, he suggested, even government will find it difficult to control the direction and outcomes of science.[22]

In fact, government-funded science is part of the problem. Although government supports much of the research in environmental and public health science, much governmental funding also goes into the development of weapons systems and into subsidies for unsustainable industrial programs, such as mining on public lands, and agricultural programs, such as the growing of sugar in Florida and corn for ethanol.

Nor is government guiltless of unethical uses of science: witness the experiments by the U.S. Army on the effects of radiation on soldiers made without the soldiers' knowledge or the experiments by public health officials on the effects of venereal disease performed on African American men.

As Milbrath cautioned, "Allowing science to develop without constraint would likely lead to a future society (50 to 100 years from now) that would be densely populated, nearly all its resources would be devoted to human purposes, wilderness and wildlife would be in sharp decline, and the biosphere would be in great danger."[23]

It would seem, therefore, that some more rational, ethical system of governance is needed to tame the genie. In our society, such governance would need to assume a democratic shape.

Democratizing Science

A document called "The Heidelberg Appeal," first publicly released at the 1992 Earth Summit in Rio de Janeiro, described itself as "a quiet call for reason and a recognition of scientific progress as the solution to, not the cause of, the health and environmental problems that we face." The appeal was signed by several thousand scientists and other intellectuals, some of them Nobel Prize winners and only a few of them recognizable as part of the "brownlash." The statement asserted that the signatories wanted to contribute to preserving the earth but went on to say, "We are, however, worried at the dawn of the twenty-first century, at the emergence of an irrational ideology which is opposed to scientific and industrial progress and impedes economic and social development." In words that could have been written by Francis Bacon more than four hundred years earlier, it declared, "Humanity had always progressed by increasingly harnessing Nature to its needs and not the reverse." The appeal subscribed to the "objectives of a scientific ecology for a universe whose resources must be taken stock of, monitored and preserved. But we herewith demand that this stock-taking, monitoring and preservation be founded on scientific criteria and not on irrational preconceptions."[24]

The appeal made no mention of global warming or the rapid loss of biological diversity, two of the key issues on the summit's agenda, nor any other damage to the environment and human welfare caused by the unintended and unforeseen consequences of science and technology as it harnessed nature.

"The Heidelberg Appeal" is a profound misreading of what environmentalism is about. Except for a radical fringe, environmentalists are by no means anti-science or anti-technology. Nor are they irrational. Environmentalism is founded on science and is driven by science. Its goal is to correct damage to the natural environment caused by inappropriate, misapplied, and unnecessary technology and to persuade scientists to produce results that will protect and enhance the natural world, protect human welfare, and provide for an economy that is sustainable into future generations. Most mainstream environmentalists see science and technology as indispensable tools for achieving those goals. But they also insist that the tools be carefully studied for their long-term effects and insist on replacement or adjustment of science and technology that cause problems or are not designed to meet the goals of a safe and sustainable environment.

There are many in the scientific community such as Jane Lubchenco who understand environmentalism and its significance and the need for broad scientific reform. The content and tone of "The Heidelberg Appeal," however, suggest that there are also many scientists who still look myopically out at the world from their laboratory windows, who do not understand that "harnessing nature" to human needs has already in some ways gone too far for the good of humanity or that, given the second law of thermodynamics, the inevitable end result of too much harnessing of nature is the acceleration of entropy. It suggests that there are still members of the scientific community who do not want to believe that the results of misconceived or misused science and technology can be social disruption, ecological devastation, disease, and death, who pretend not to see that much of science and technology is directed by those with money and power, those who, seeking more money and power, will sponsor science and technology that does not meet real human needs. It suggests that there remain many scientists who still, in the face of all evidence to the contrary, regard science as the only source of the True and the Good.

The real message of "The Heidelberg Appeal" is a far cry from the spirit of Lubchenco's proposed social contract. It is a group of scientists saying to us, "In spite of the harm some of our work has done and is doing, trust us and leave us alone."

But we can no longer do that. Given the gravity of threats to the biosphere, we can no longer afford to accept the bad along with the good that science gives us. The power of science and technology, their key role in determining our future, is manifestly too important and crucial to leave solely in the hands of scientists and engineers and those who pay for science and technology. What is needed is more societal control of science and technology, more *democracy* in determining scientific priorities and in judging which uses of science serve the welfare of humanity and its habitat and which of them might endanger us. All of us have a vital stake in the outcome of the scientific enterprise; all of us should have the opportunity to help guide it. As Lester Milbrath stated: "We should be wary of the notion that scientists know better than the people themselves what is good for the people. Science policy making should be in an open enough forum, and be given sufficient time, so that citizens can meaningfully play a role in the development of policy." [25]

Such social control is not a new idea. A quarter of a century ago, the administration of President Gerald Ford looked into the idea of "science courts" as a means of resolving disputes over science. Shortly thereafter, Sheldon Krimsky suggested in the *Bulletin of the Atomic Scientists* the creation of "citizen courts" to be appointed by local governments to evaluate and help steer science policy. Milbrath has proposed a "Council for Long-Range Societal Guidance" that would facilitate social learning about scientific issues and other issues of sustainability and would conduct assessments of the long-range effects of contemplated policy.[26]

In recent years, the notion of "civic science" or "postnormal science" has gained a foothold among some members of the scientific and policy-making communities. As Kai N. Lee, former director of the Center for Environmental Studies at Williams College, described it, civic science "should be irreducibly public in the way responsibilities are exercised, intrinsically technical, and open to learning from errors and profiting from successes."[27] Lee proposed that society find its way to appropriate scientific policy through a long-term process of trial-and-error learning, which he called adaptive management. To enable us to steer the proper course through this process, he said, we can make use of two powerful navigational aids, the "compass" of the physical and social sciences and their rigorous disciplines, to point us in the right direction, and the "gyroscope" of democratic debate within the political process, in which the competition of interests and ideas can keep science from veering off the course it must set to extract us from our ecological predicament.

In an essay titled "A New Scientific Methodology for Global Environmental Issues," systems engineers and researchers Silvio O. Funtowicz and Jerome R. Ravetz argued: "We have now reached the point where a narrow scientific tradition is no longer appropriate to our needs. Unless we find a way of enriching our science to include practice, we will fail to create methods for coping with the environmental challenges, in all their complexity, variability and uncertainty."[28] Their solution is the establishment of "postnormal" science that would open the scientific process to vigorous, open debate. It is postnormal, they explained, in contrast to "normal" science, in which all outsiders are excluded from the dialog among the initiates of a particular scientific discipline. "But since the insiders are manifestly incapable of providing effective conclusive answers to many of the problems they confront, the outsiders are capable of forcing their way into a dialogue," they noted. "When the debate is conducted before a lay public, the outsiders (including community activists, lawyers, legislators and journalists) may on occasion even set the agenda." These "extended peer communities," Funtowicz and Ravetz argued, are crucial in enabling science to respond to the challenges of global environmental issues. "The democratization of political life is now commonplace; its hazards are accepted as a small price to pay. Now it becomes possible to achieve a parallel democratization of knowledge, not merely in mass education but in enhanced participation in decision making for common problems."[29]

One proposal for a more democratized environmental science is being pushed by the Committee for the National Institute for the Environment (CNIE), a group headed by former U.S. ambassador Richard E. Benedick. The CNIE's goals are to establish a National Institute for the Environment to gather and disseminate information on environmental science, create a National Library for the Environment, and conduct a two-way dialog with the public on relevant issues.

In recent years, there has been a rapid expansion of the tools needed for wide public participation in scientific decision making. The Internet and the World Wide Web, of course, are the most dramatic examples of the exploding information base available to all citizens. Right-to-know laws such as the Toxics Release Inventory are giving people data that in the past were easily accessible only for government regulators, members of the scientific community, and those who paid for science and technology.

The idea of having democratized science is not to interfere in basic research or hamper scientific and technological creativity; it is not to peer over the shoulders of scientists seated at their microscopes or retorts. The idea is to have a broadly representative segment of society participate in decisions about the direction and deployment of science and technology to help ensure that science serves the broadest public good and does not lead us down paths that threaten our present and our future.

The Precautionary Principle

Physicians today continue to feel themselves bound, at least theoretically, by Hippocrates' definition of their role some 2,400 years ago: "to help, or at least, to do no harm." The practices of science and technology, however, function under no such constraint. Science habitually pursues knowledge and utility with scant regard for ultimate consequences. Technology is often, perhaps usually, deployed in the marketplace with inadequate investigation into potential long-term effects on human health and the environment—with assurances that it will do no harm. For example, there are more than 70,000 chemicals in commerce in the United States, only a tiny fraction of which have been fully tested for their ability to cause harm to health and the environment.[30]

In the United States, at least, the burden of proof for showing that the result of science and technology is or can be harm to society or individuals falls not on scientists, engineers, or those who commercialize their work but on those suffering the harm. Governmental agencies such as the Environmental Protection Agency, the Food and Drug Administration, and the Federal Trade Commission are overwhelmed in their efforts to protect the public by the sheer volume of industrial and commercial activity accompanying the marketing of the products of science and technology. Industrial codes and standards are a flimsy shield. Individual citizens must rely on tort law, to which they can resort only after they have suffered injury. Wildlife and natural beauty do not have even that means of redress.

There is, however, a potentially powerful weapon that could be used in democratic oversight of the deployment of the products of science and technology. It is called the precautionary principle, and it is being regarded with increasing interest by environmentalists as well as some policy makers, members of the scientific community, and even some businesses and industries.

The concept is a simple one. In essence, it says "better safe than sorry" when making a decision about exposing people or the environment to a new technology, product, process, or substance. It would shift the burden of proof to those who invent, produce, and distribute products or technologies, or who advocate their use, to demonstrate that they will do no harm over their entire life cycle. If clear proof can be presented, then the technologies can be deployed. Conversely, in the absence of compelling scientific evidence, the precautionary principle would preclude the use of products or technology. Similarly, when confronted with an apparently grave existing threat such as a change in the earth's climate, the precautionary principle would require action to mitigate the threat even if the science is not yet conclusive but is supported by the weight of evidence.

The precautionary principle gives due regard to the welfare of future generations as well as that of the current generation. Economist Charles Perrings explained, "This principle can be interpreted as saying that if it is known that an action may cause profound and irreversible environmental change which permanently reduces the welfare of future generations, but the probability of such damage is not known, then it is inequitable to act as if the probability is known."[31]

Use of the precautionary principle would be initiated by scientists, engineers, and technicians, who would be under professional obligation to withhold their work from commerce until they are satisfied it will cause no harm. The next line of defense would be manufacturers, distributors, and marketers, who would be faced with *caveat vendor* rules about the effects of their products on the public and the environment.

But final decisions about technologies and their products would be made by extended peer groups, including communities that could be at risk. Decisions would be based on careful analysis, review, and dialog. Broader participation would bring to the process new perspectives and broader, perhaps decisive, information and wisdom. For example, a British survey indicated that sheep farmers in Cumbria sometimes had a better understanding of the ecological effects of radioactive contamination from nuclear power plants than did scientists.[32]

Although the precautionary principle is intended to prevent harm, it could help ease and accelerate public acceptance of otherwise suspect technologies. A case in point is the current deep unease over the use of genetically engineered products, particularly food products. Genetic engineering could prove a substantial boon to humanity by addressing serious problems such as predicted food shortages in the face of continuing population growth. But because of uncertainties regarding possible disastrous effects, the technology is widely resisted, and in some parts of the world, genetically altered products are banned outright. Governmental agencies such as the Food and Drug Administration are an important line of defense, but they do not have the resources to review adequately the torrent of products coming to the market. Nor can they be completely free from political pressures. However, if technologies and products were subjected to highly public peer review extending beyond governmental and marketplace oversight and then given a seal of approval, it is likely that much of the fear and suspicion would dissipate and their path to the market would be smoothed.

Environmentalists and Science

Environmental regulation has been described as "a shotgun marriage of science and law."[33] Environmentalism as a movement is a shotgun marriage of science to just about everything—science and politics, science and economics, science and justice (which, alas, is something other than law), science and religion, science and ethics, science and aesthetics. Environmentalists find out what they must do from science and technology—for example, learning from scientists such as Sherwood Rowland and Mario Molina, who won the Nobel Prize after discovering the potential destruction of the earth's ozone shield by the widely used industrial chemicals called chlorofluorocarbons (CFCs)—and then they urge action based on the scientific solution, banning, for example, the production and use of CFCs and related compounds. One of the chief roles of environmentalists is to act as intermediaries between science and the public, the media, and lawmakers.

Given the centrality of science to environmentalism, the scientific capacity of our environmental organizations is clearly inadequate. Henry W. Kendall, chairman and longtime guiding spirit of the Union of Concerned Scientists, said shortly before his death that "the environmental community's understanding and use of relevant science is poor. They don't even understand the potential limitations of science. So they can't distinguish between legitimate science and what is science fiction or fantasy." That was probably an excessively severe assessment. There are a number of first-class scientists within the environmental movement, and they are listened to with respect by the scientific community. But there is no question that the environmental movement needs to expand and upgrade its scientific capacity—not necessarily for basic research and development but for a wide range of other tasks. Just as environmentalists need increased political and economic power to achieve their goals, they require more of the power conveyed by scientific expertise.

If science is to be democratized, if future decisions about priorities for and deployment of science and technology are to made by broad civic participation, environmentalists will need the knowledge and capacity to be represented—and be effective—at the table. They will need resources to monitor and assess new science and new and existing technology. And not only the national environmental organizations will require that capacity; if local communities are to participate in the decision-making process, they, too, will need to bring a degree of expertise to the table. Here, perhaps, is another area in which national groups can help the grass roots with information, technical assistance, and training—provided the nationals themselves first acquire sufficient resources. If, for example, members of a community fear that a planned toxic waste facility will harm their children, a national organization with sufficient scientific capacity can provide data and analytical tools to help the community make its case and can train local activists to gather data and make their own analyses.

Environmentalists also need to substantially increase their ability to translate scientific problems and scientific solutions and communicate them to the public to make clear what the real stakes are. Dr. Eric Chivian, director of the Center for Health and the Global Environment at Harvard Medical School and an activist with Physicians for Social Responsibility, said he thinks that one of the greatest contributions of physicians has been "to humanize and personalize the threat of nuclear weapons." Similarly, environmentalists need to do a much better job of personalizing and humanizing environmental issues for the American people. As the scientific database grows ever more complex and its language ever more arcane, there is an urgent need to make it more accessible to the media, the public, and policy makers. Environmental organizations can help do that by working with existing scientific institutions and associations and by promoting new sources of public access to science, such as the proposed National Institute for the Environment.

Most urgently, environmentalists need to throw their full weight into what will be a long, arduous campaign to shift the very course of the scientific enterprise toward the path suggested by Jane Lubchenco's social contract. In their current mode, environmentalists seek to remedy one at a time the damages to the environment caused by the mistakes and unforeseen consequences of science. But as Lynton Caldwell cautioned, "It seems apparent that the integrity of the biosphere and mankind's ecological and economic safety on Earth will not be secured by incremental measures alone."[34]

"It's a scary time," said George Woodwell, director of the Woods Hole Research Center. "We know the world is not working. It is up to science to define what will work in a biophysical sense." He added, however: "I look at the scientific community and see it isn't really structured to live up to the challenges . . . as far as getting together on the great issues on a sustained basis, they just aren't there. So I look to the environmental community for innovations and ideas, the recognition of the seriousness of the problems and their solutions. But it is a small community, way too small and beleaguered."

To influence the course of science and technology in the 21st century, the environmental movement will have to grow in size and capacity—and become the beleaguerers instead of the beleaguered.

Chapter 8
Small World: America and the Global Environment

As I write this chapter, hundreds of thousands of ethnic Albanians are fleeing for their lives from Kosovo while missiles and bombs from planes and warships of the North Atlantic Treaty Organization rain down on wide areas of Yugoslavia. Bombs have again been exploding in Northern Ireland despite the painstakingly negotiated agreement between Protestant and Catholic militants. A stable peace in the Middle East seems as elusive as ever. Ethnic, tribal, and religious hatreds continue to produce tragic bloodletting in parts of Africa and South Asia. Poverty and gross economic inequity have led to armed uprisings and murder in several Latin American countries. One in every five humans remains in deep poverty.

Less than a decade ago, just after the end of the cold war, the heads of most of the world's nation-states gathered in Rio de Janeiro to plan a new way for the human community to live on this planet. The goal of the Earth Summit was to set new priorities for peoples and governments in order to plan a future of ecologically sustainable and socially equitable economic development. The summit produced a massive document called Agenda 21, which was supposed to serve as a road map to a postindustrial, postmodern era of economic, political, and environmental sanity. Instead of the bipolar, armed confrontation between superpowers that governed geopolitics for decades, cooperation among nations to preserve the planet and its people would be the unifying theme of a rapidly globalizing society in the 21st century. Care for our shared habitat would be the basis of a new *Pax Gaia*—Peace of the Earth.[1]

It has not quite worked out that way. Globalization is proceeding at astonishing speed, all right, but at the turn of the century, there is no *Pax Gaia* in sight.

New York Times columnist Thomas L. Friedman observed that the cold-war system already has been replaced by globalization, which involves "the integration of free markets, nation-states and information technologies to a degree never before witnessed, in a way that is enabling individuals, corporations and countries to reach around the world farther, faster, deeper and cheaper than ever. It is also producing a powerful backlash from those brutalized or left behind."[2]

Indeed, looking at the Balkans, Africa, and other bleeding and troubled parts of the world, it seems as if large parts of the human community, far from seeking to become part of a global community, are retreating to an earlier period of social evolution marked by endless ethnic, religious, and territorial strife. Some scholars foresee the globalization of the economy, the spreading power of global corporations, and the communications revolution of the Internet leading to disintegration of the nation-state and a return to something like medievalism, with the global economy assuming the role played by the church in the Middle Ages.[3]

Meanwhile, there has been little progress toward the ideal of sustainable development—toward integrating economic progress and equity with protection and restoration of the global environment. Maurice Strong, secretary general of the Earth Summit, said: "After Rio, we went into environmental recession. Governments haven't changed their rhetoric, but their performance has weakened. . . . Agenda 21 is a set of benchmarks against which our lack of progress is being measured."

Instead of making progress, the global community has been in retreat in some of its most vital commitments to achieve a secure global environment and to reduce global poverty and inequity. Threats to the global environment such as rising carbon dioxide levels, deforestation, and loss of species continue to widen. Instead of the richer countries increasing their development assistance to the poorer countries to help them attain environmentally sustainable economic development, such assistance has actually declined—instead of the promised 0.7 percent of gross domestic product for aid, the rich countries are providing, on average, less than 0.3 percent. The U.S. contribution has been less than 0.2 percent in recent years as conservative Congresses have reduced aid budgets. Some of the opposition to foreign aid was based on the belief that much of it was being misused by corrupt governments. Private investment in developing countries has exploded, but it is largely directed at projects intended solely to produce profit, and unguided by national or international policy makers, it is not designed for ecologically healthy development that will move people from poverty and end pressure on their resource base. The income gap between rich and poor is still widening.[4]

Nearly ten years after Rio, it is painfully apparent that governments and international institutions that are not acting or that are acting too slowly and inadequately must be vigorously prodded. If they will not act, others must act in their place. Because environmentalism is one of the biggest, most widely accepted, and most sophisticated independent sector movements, it must do much of the prodding and take the lead in generating action. Now, at the beginning of the 21st century, is the time when these global ecological issues must be fully addressed. As Henry W. Kendall stated, "The magnitude and number of challenges and the speed of their arrival are unlike any that human beings have faced. Little time remains—so little that it is our generation that must face and resolve these challenges."[5] Or, as the 1995 report of the Commission on Global Governance put it, we must have "better management for survival."[6]

Since Rio, it has become clear—if ever it were not—that the global environment cannot be rescued and secured in isolation from the nature and condition of the global economy, international trade, geopolitics and global governance, and the social conditions into which the people of the world are bound. Environmentalists in the United States and around the world will have to adjust more rapidly than their governments to rapid globalization and to the centrifugal forces that are pulling apart areas such as the Balkans. Because the United States is the world's only superpower and because it is the most powerful engine driving the global economy, it will, by default, have to set an example in mending the economic, political, and social flaws that are leading to the erosion of global systems at such a frightening pace. Given the country's retrograde politics of recent years and its fervent commitment to market absolutism at the end of the century, the American environmental movement will have to provide a stronger conscience and a louder voice to push the United States into becoming the guardian rather than the assailant of the global environment.

Global Corporations and the Global Economy

The great multinational corporations—perhaps *global* corporation is now a more descriptive term—many of them more rich and powerful than the majority of sovereign states, are responsible for many of these flaws and have the capacity to correct them. As Stephen Viederman noted, half of the one hundred biggest economies in the world are corporate. Companies such as the General Motors Corporation, the Exxon Corporation, the Mitsubishi Corporation, and others have annual revenues exceeding the gross national products of most nation-states. Increasingly conducting their production and marketing operations in all countries, they are effectively restricted by the laws of no country—nor is there a body of hard international commercial law to control their activities.

Political scientist Benjamin R. Barber found that at the end of the 20th century, two very different and antagonistic forces were tending to erode the authority of nation-states. One was the rise of ethnic militancy and fanatic religious fundamentalism, for which he uses the term *Jihad,* the Arabic word for holy war. The other was the growing power of the multinational corporations and the globalization not only of the economy but also of culture and values, a force Barber calls "McWorld," with a nod toward the mushroom-like emergence of the golden arches of McDonald's Corporation franchises in every corner of the planet. Barber believes that over the long run, McWorld, which serves "corporate welfare, not the public good," appears to present the greatest threat to efforts to protect the environment, workers, and democracy.[7] "Markets," he wrote, "have emerged triumphant from a war against the nation-state and the public interests they represent that has been waged at least since Adam Smith." In the name of economic globalization, he added, environmental and employment policies and even democracy are regarded as "narrow interests."[8]

In late 1998, the International Forum on Globalization, an alliance of environmentalists, scientists, economists, and human rights activists, published a full-page advertisement in the *New York Times* (and presumably in other media) warning: "All peoples of the world have been made tragically dependent upon the arbitrary, self interested acts of giant corporations, bankers and speculators. This is the result of global rules that remove real economic power from nations, communities and citizen democracies, while giving new powers to corporate and financial speculators." The statement also asserted: "No system that depends for its success on a never ending expansion of markets, resources and consumers, or that fails to achieve social equity and meaningful livelihood for all people on the planet, can hope to survive for very long. *Social unrest, economic and ecological breakdown are the true inevitabilities of such a system*" (emphasis in original).[9]

Examples abound. In the mid-1990s, a unit of Daishowa Inc., a Japanese conglomerate, obtained rights from the government of the Province of Alberta, Canada, to log 4,000 acres inhabited by members of the Lubicon Lake Cree. When environmentalists and human rights activists organized a boycott of Daishowa products to try block clear-cutting of the territory, the company sued, and an appellate court ordered the boycott to end. The decision was widely viewed by Canadians as a successful effort by a multinational corporation to silence the dissent of citizens in their own country.[10]

Although there is widening agreement that economic globalization is eroding the authority of national governments, many observers see that as an irreversible trend that will not necessarily be bad for the cause of protecting the environment or the broader cause of enhancing human welfare. Harvard economist Dale Jorgenson said that globalization "by its nature" will erode the nation-state, but that simply means "we are going to have to rely much more on the global economy and less on our national government to reach our social goals." David Buzzelli, former cochair of the President's Council on Sustainable Development and former vice president and corporate director for environment, health, and safety of the Dow Chemical Company, stated: "Globalization is for the most part positive. Companies that operate around the world in many different countries have a tendency to adopt the best [environmental and labor] practices and spread them around the world. Globalization is raising the bar of what is happening, particularly in the developing countries." Maurice Strong, a successful businessman as well as international public servant, thinks that the "denationalization" of corporations and their ability to move money and experience rapidly across borders will produce a "chaotic world." But he added that global corporations will cooperate "to help save the world because it is in their interest. They cannot operate in chaotic markets. They have a vested interest in a reasonably secure and functional world." Paul Hawken noted that, as the Shell Oil Company found out when it wanted to sink an unwanted offshore oil rig into the North Sea but abandoned the plan after a public outcry provoked by environmentalists, "globalization makes companies much more vulnerable to pressure such as boycotts, bad publicity, and twists of fate. Their soft belly is the Internet, which will make companies more accountable in more places than they possibly could have imagined."

There are only scattered signs, however, that corporations are acting or will act to create a sustainable and equitable global economy. Business professor Stuart Hart, in an article in the *Harvard Business Review,* asserted that the corporate world has the responsibility and the capability to build a sustainable global economy and that it makes "good business sense to do so." Currently however, few companies have incorporated sustainability into their strategic thinking, focusing on only a few piecemeal projects to control pollution. He contended that corporations should not push for rapid growth in emerging economies because such growth cannot be sustained "in the face of mounting environmental deterioration, poverty and resource depletion."[11] In another article, Hart insists that global sustainability could be "the biggest commercial opportunity in recent history."[12]

Unfortunately, however, the global economy being constructed with breathtaking swiftness by the multinational corporations and financial institutions still is based on the unsustainable McWorld model that characterizes the high-growth, high-consumption, energy-intensive economies of North America, western Europe, and Japan. Such an economy is like someone on a treadmill that goes faster and faster, uncontrolled by the runner, who cannot get off without badly injuring himself.

A case in point is the Asian financial crisis of the late 1990s. The crisis ended a period of feverish economic growth by Thailand, Indonesia, Malaysia, and other developing countries that was fueled by heavy speculative investment from Japan, Europe, and North America. When foreign capital started to flee from the Asian countries in 1997 following a series of corporate bankruptcies and bank failures, the economies of those countries went into deep tailspins, leaving behind millions of unemployed workers and hungry families and growing pressure on natural resources such as timber as governments scrambled for ways to meet their debts.[13]

An economy like this cannot be sustained indefinitely even in the wealthy developed countries. But the peoples of the world who have not yet acquired such economies, prodded by the relentless and persuasive marketing and advertising of the McWorld corporations and enticed by the increasingly globalized information and entertainment industries, want to achieve it. And so will the additional 3 billion to 5 billion people who will be born into the human community over the coming decades. No less an authority than Robert Shapiro, chief executive officer of the Monsanto Company, is convinced that if those additional people "follow the same path to improving their lives as the people did from the beginning of the 19th century until now, the world is over. It cannot be done. The planet cannot support the consequences of another industrial revolution."

On the other hand, there is unlikely to be a commitment to protecting the global environment in the absence of growth in the developing and emerging economies. Lee Thomas, former administrator of the Environmental Protection Agency, noted: "It is a chicken-and-egg thing. In China, for example, there is heavy industrialization to improve the economy, utilities, and infrastructure without concern for local and global environments, and that will cause problems. But without basic economic improvement, [the Chinese government] will not put a priority on the environment."

There is a wealth of ideas for achieving both goals without breaking the egg. That is the meaning of sustainable development. It is the long-term strategy described in Agenda 21 and other detailed prescriptions such as Caring for the Earth, a program proposed in 1991 by the World Conservation Union (IUCN), the United Nations Environment Programme, and the World Wide Fund for Nature.[14] Economic growth can be linked with environmental protection and social justice. Achieving such an economy will take time, experimentation, and what Kai Lee calls "adaptive management"—finding out which policies and programs work and which do not and then adjusting.[15]

Unfortunately, as Sir Martin Holdgate, former executive director of the IUCN and now president of the Zoological Society of London, lamented, "There is an immense gap between strategy and action."[16] In today's rapidly globalizing economy, which is being shaped by Darwinian competition among giant corporations unaccountable to national laws and politics as well as tsunamis of private capital rushing unimpeded across national borders, there appears to be little will to create a sustainable economy and little authority to compel it. "Development," noted Jonathan Lash, president of the World Resources Institute and, with Interface, Inc.'s Ray Anderson, cochair of the President's Council on Sustainable Development, "is no longer controlled by international institutions because it is overwhelmed by private investment. The problem is, how do you protect environment when it is not protected by market pricing? How do you protect the oceans, the river basins, the soil, and assure maximum market efficiency at the same time? How do you deal with corruption in the marketplace?"

Economist Herman Daly regards the late-20th-century "rush to globalization" as the "last gasp of the growth economy." Having saturated their own markets, he argues, the developed countries are "growing into the commons and into each other's economies." Globalization, he warned, "is a major threat to environmental improvement. As the economy globalizes, nations lose their ability to carry out other policy for the national good."

A globalized economy, however, is likely to be a fact of life that environmentalists will have to deal with in the 21st century. Given the reach and control of the global corporations, the enormous physical and financial resources at their command, the networks of economic transactions that cross national borders with impunity, and the absence of legal impediments, it is difficult to see how the current system will exit the world stage. The environmental movement will have to deal with globalization and find ways to make it an ally rather than an antagonist of environmental protection. One key area environmentalists will have to address is the drive to eliminate trade barriers and harmonize international rules that characterized the latter part of the 20th century.

Trade and the Environment

In 1998, the Canadian government reversed a ban it had imposed on the gasoline additive MMT, which it had found the year before to be a pollutant and a health hazard, after the additive's U.S. manufacturer, the Ethyl Corporation, filed a lawsuit claiming the ban violated the North American Free Trade Agreement (NAFTA). Canada's decision came shortly after the World Trade Organization overruled a U.S. law that required any shrimp imported into the United States to be caught with nets that use special devices to keep sea turtles, a number of species of which are in danger of extinction, from being caught and drowned in the nets.[17]

These decisions confirmed the fears of many environmentalists who believed that the drive to remove trade barriers and integrate economies would lead to erosion of national environmental laws and "harmonization" of environmental laws at the lowest common denominator of environmental protection—either deliberately or as an unintended side effect.

Trading of goods and services has been going on since prehistoric times, and trade across national borders has been taking place since there have been nations. As environmental lawyers Durwood Zaelke, Paul Orbuch, and Robert Housman pointed out, "Trade has played a major role in world progress, bringing pasta to Italy, silk to France, Columbus to the Americas, polio vaccine to the world, and Ben & Jerry's ice cream to Moscow."[18] Trade has helped create wealth, jobs, and prosperity in the developed countries of the North. It is considered vital to the economic development of the poorer countries—although often the terms of trade are stacked against them and can weaken rather than strengthen their economies. Trade and economic integration are also regarded as instruments for promoting peaceful relations among nations. The European Union was born after World War II because leaders and peoples of that bloody continent were casting about desperately for ways to end the barbarous cycle of violence that had claimed so many millions of lives and destroyed so much of its civilization in the 20th century.

Liberalized trade and economic integration have also caused hardships for companies that do not operate as efficiently as competitors in other lands, for workers who lose their jobs to workers in other countries who are paid lower wages and receive fewer or no benefits, and to farmers who see their domestic markets shrinking before a flood of cheap imported food. Traditionally, those affected negatively by foreign trade—American steelmakers, French wheat farmers, Japanese rice growers, and countless others—have demanded protection from their national governments and have often received it in the form of tariffs or, in the case of some countries, such as Japan, in both tariffs and an artful web of nontariff barriers. In recent years, however, as economic globalization has accelerated and trade blocs have been formed, protective barriers have been lowered or removed.

In the past, environmentalists generally sat on the sidelines in trade debates. If anything, they supported freer trade on the ground that it would improve economies in the developing countries and put them in a better position to protect their own resource bases. Toward the end of the 20th century, however, environmental attitudes began to shift. In the debate over NAFTA, the American environmental movement was divided virtually down the middle over whether the United States should join the trade pact. Some of the national and international groups continued to support the elimination of trade barriers. Others, fearing that Mexico's weak enforcement of environmental laws would lead to a buildup of polluting industries along its border with the United States and that American environmental laws would be overridden by trade rules, joined with labor unions in opposing the treaty. Although support for NAFTA from some of the mainstream groups helped in getting a Commission for Environmental Cooperation into the treaty, many environmentalists, including some who supported it originally, think that environmental considerations are given second-class status in decisions by administrators of the pact.

Environmentalists regard the World Trade Organization with even more suspicion because of its propensity for riding roughshod over environmental needs as well as labor and social values in a single-minded drive to smooth the way for international trade. Elizabeth Dowdeswell, former executive director of the United Nations Environment Programme, and Steve Charnovitz, director of the Global Environment and Trade Study, a research program of the Yale Center for Environmental Law and Policy, observed: "Neither the postwar trade negotiations known as the General Agreement on Tariffs and Trade (GATT) nor its successor body, the World Trade Organization (WTO), has made significant progress in integrating environmental considerations into the trade domain." They noted that there is a danger of environmental policy being manipulated for protectionist purposes but added, "Nations are beginning to realize that optimal trade policies cannot be set without taking environmental effects into consideration and vice versa."[19]

Some observers, including Harvard economist Dale Jorgenson, believe that trade and environmental policy should proceed on separate tracks. "We can't tell Mexico or Thailand or Uruguay what health policies to pursue in their own country. So it is better to set trade policies that are independent of public health and environmental standards," Jorgenson contended. But this appears to be increasingly a minority view as more and more scholars and policy makers are recognizing an inevitable link between global trade and the environment. Daniel C. Esty, director of the Yale Center for Environmental Law and Policy, noted: "The power of economic integration is very great—there are hard economic drivers behind it. There is money to be made, and the average person stands to benefit from lower prices and a greater variety of goods. Steering this force is what the World Trade Organization is about. They have to set bounds of behavior, and one is [that] you don't foul the environment. But environmental ministries are generally subordinate to economic and trade ministries, and WTO rules are slanted to promote trade over the environment. They have displayed a shocking failure to learn the lesson that optimal trade policy is made by including the environment, not by ignoring it."

Environmentalists also complain that the procedures for setting trade policy are undemocratic. Although in the United States, trade treaties must by ratified by the Senate, they are implemented, particularly within the World Trade Organization, by international bureaucracies that are not held accountable and that often conduct their deliberations in secret, without input from the public. For example, three anonymous officials, deliberating behind closed doors in Geneva, ruled that a U.S. law barring importation of tuna caught by methods that killed dolphins was in violation of the General Agreement on Tariffs and Trade. Environmental policy, in contrast, is almost always made under the full light of public scrutiny and usually with wide participation by corporations, environmentalists, and other nongovernmental organizations.

But there is no inherent conflict between free trade and adherence to global environmental standards. Provided the standards are not abused for covert protectionism, there is no reason why international economic activity should not be constrained by the same considerations that guide environmental policy in a growing number of countries. Nor is there any reason why an organization such as the World Trade Organization should not operate in the full sunlight of open, democratic participation. Rules, procedures, and institutions can be changed and improved.

As Dowdeswell and Charnovitz observed, "Sustainable development requires a comprehensive perspective that integrates environmental, social and economic goals. Governments should be prepared to introduce reforms that recognize the new economic realities and address past international policy failures."[20] To do so, however, governments must have a workable system of international governance with binding rules and institutions to carry them out. There is not yet such a workable system.

Governing the Global Commons

The latter half of the 20th century witnessed not only recognition of the grave dangers to the planet's physical integrity being created by human activity but also a remarkable burst of scientific and diplomatic effort to construct an international response to those dangers. Since the United Nations Conference on the Human Environment, held in Stockholm in 1972, national governments have negotiated nearly 100 multilateral treaties and conventions to protect the global environment. New institutions have been formed to monitor the global environment and address threats to its integrity, including the United Nations Environment Programme, created after Stockholm, and the United Nations Commission on Sustainable Development, inaugurated after the 1992 Earth Summit in Rio de Janeiro. As the cold war ebbed, environmental concerns, previously on the far periphery of geopolitics, moved toward its center.

However, although geopolitics may be greening, many believe it is doing so too slowly to respond to the grave and rapidly mounting threats to the global environment. Moreover, as pointed out by Gareth Porter and Janet Welsh Brown, there is a fatal flaw in the current international effort to meet the global environmental challenge: "the absence of an effective enforcement mechanism."[21] Lynton Caldwell observed that "governments have found it easier to sign declarations and collaborate in joint scientific investigations . . . than to fulfill environmental agreements through regulatory measures of their own, or through conformity to international policies and standards."[22]

The United Nations, formed in 1945 as World War II was ending, is the closest thing we have to an institution of international governance. The United Nations was, in theory, created to serve the people of the world, to maintain international peace and security, and to act as an agency of international cooperation to achieve common goals for the good of humanity. In practice, however, it has been unable to live up to the vision of its creators. For much of its history, it has served as the chief diplomatic venue for the hostile East–West confrontation between superpowers and as a forum for airing the grievances of the poorer and less powerful nations in the long-running argument between North and South. Many parts of the complex structure of the United Nations have become self-absorbed, self-serving bureaucracies, ossified in minutiae and forgetful of their role as servants of what the United Nations charter calls "We the Peoples." A 1995 report by the Commission on Global Governance stated: "Fifty Years after [its founding in] San Francisco, the United Nations is viewed predominantly, by both people and governments, as a global third party—belonging to itself, owned by no one except its own officials, and even, to an extent, dispensable."[23]

That is not to say that the United Nations is a black-and-white illustration of failure. It has put in place an extensive foundation of international mechanisms to address a variety of human needs, including economic development, human rights, rescue of refugees, the welfare of children, improvement of human health, promotion of food production and distribution, advancement of more efficient international transportation and communication, improvement of human habitats, and cooperation in scientific discovery. With all its inadequacies, it is still the chief instrument of international cooperation on the environment. It has the potential to be much more.

As now constituted, however, the United Nations cannot be the political driver of efforts to protect the global environment. It was not created as an instrument of governance. The powers that called the United Nations into being were careful not to give it the authority to limit their sovereignty. The United Nations Environment Programme, for example, has no means of compelling states or their citizens to refrain from polluting or depleting the global commons. Irish diplomat and historian Conor Cruise O'Brien asserted, with only a modicum of hyperbole: "The United Nations cannot do anything, and never could. It is not an animate entity or agent. It is a place, a stage, a forum and a shrine . . . a place to which powerful people can repair when they are fearful about the course on which their own rhetoric seems to be propelling them."[24]

What now passes for global governance therefore depends on the willingness of national governments to pool their sovereignty, to adhere voluntarily to international laws and treaties, and to support the decisions of international institutions. International authorities such as the World Trade Organization and the European Union are becoming more prominent actors on the global stage, but as the Sierra Club's Carl Pope noted, national governments still write the script. Many nations—none more so than the United States—still refuse to share sovereignty to any substantial degree or to give due support to international institutions. The U.S. government has long been in arrears in its financial obligations to the United Nations—in 1999, it owed roughly $1 billion, including expenses associated with peacekeeping operations. The United States often has to be dragged kicking and screaming into international treaties to protect the environment—for example, by late 1999, President Clinton had not submitted the Kyoto Protocol for dealing with global warming to the Senate for ratification because of the certainty that it would be rejected. In the name of what it deems to be the national interest and freedom of the marketplace, the United States has been a increasingly reluctant global citizen. The estimable Daniel Patrick Moynihan noted that "for the longest while the United States professed a strong attachment to the idea of law in the relations of states" but added that "somewhere along the line, this conviction faltered." The result, he said, has not been good for this country,[25] particularly given the emergence of "new-order" international issues such as weapons control, human rights, and "environmental concerns that obliterate borders."[26]

Just so. Global environmental problems make a mockery of any nation's claim of full sovereignty. As Benjamin Barber observed, "even the most developed country cannot pretend to genuine sovereignty in the face of ecological 'Armageddon.'"[27] There is also a fine irony in the fact that politicians in countries such as the United States who are zealous in guarding national sovereignty when it comes to international cooperation on such issues as protection of the global environment happily support the surrender of sovereignty to global corporations in the name of free-market capitalism.

It is not a comforting picture. But as the Commission on Global Governance, a panel of distinguished former national leaders, asserted, "There is no alternative to working together and using collective power to create a better world."[28] Without a workable system that enables the nations and peoples of the world to act together collectively and decisively, we will not be able to turn back the threats to global ecological systems—or to reduce armed conflict, achieve economic equity, and secure human rights for all members of the global community. The commission, which was formed in 1992 at the instigation of the former West German chancellor and global statesman Willy Brandt, proposed a broad agenda for creating a system of global governance. Among its recommendations were that the United Nations be reformed and strengthened and that its Trusteeship Council, originally chartered to oversee Trust Territories after World War II, be given authority—with teeth—to exercise stewardship over the global commons. The Hague Declaration, issued in 1989 by the prime ministers of France, the Netherlands, and Norway, called for a global environmental legislative body under the auspices of the United Nations with the power to impose environmental regulations on nation-states.[29] More recently, British diplomat Sir Crispin Tickell proposed an international "green police force" to enforce environmental laws and treaties.[30]

Perhaps the most important contribution of the Commission on Global Governance was its recognition of the need to broaden and democratize global governance: "Global governance, once viewed primarily as concerned with intergovernmental relationships, now involves not only governments and intergovernmental institutions but also non-governmental organizations (NGOs), citizens' movements, transnational corporations, academia and the mass media. The emergence of a global civil society, with many movements reinforcing a sense of human solidarity, reflects a large increase in the capacity and will of people to take control of their own lives."[31]

This is where the environmental movement comes in.

The Role of American Environmentalism

As the new century begins, the United States stands as the world's leading power, with both the biggest economy and the most powerful military. It plays a decisive role in global politics and diplomacy. If there is to be concerted, effective international cooperation to safeguard the global environment, the United States not only must be an active participant but also must set an example for the world to follow. A paramount task of American environmentalists in the coming years, therefore, will be to exercise all their skills and strength to see that the U.S. government and U.S. corporations fulfill that role.

As former Environmental Protection Agency administrator William K. Reilly noted, American environmentalists are among the few powerful advocates of an engaged foreign policy in this country's current political matrix. Fred Krupp of the Environmental Defense Fund believes that without activist American environmental scientists such as George Woodwell and Michael Oppenheimer, there would be no global warming treaty. American environmental organizations were ubiquitous throughout the entire Earth Summit process, participating actively on the U.S. delegation, lobbying at the preparatory committee meetings and at the summit itself, forming networks and devising strategies with environmentalists from other countries. They are engaged participants in many international gatherings on the environment, and several of them do a good job of educating their members and the media on global issues and their status. A number of American environmental organizations, including the World Wildlife Fund, the World Resources Institute and the Worldwatch Institute, the National Wildlife Federation, the Environmental Defense Fund, and the Natural Resources Defense Council, play important and varied roles in the international drive to deal with global environmental problems.

Many American environmentalists agree, however, that the movement has inadequate capacity to deal with global environmental issues and gives those issues far less attention than their scope and urgency require. Gene Karpinski of the U.S. Public Interest Research Group (U.S. PIRG) stated: "U.S. NGOs are not close to performing well on global issues. The community has done an okay job on the scientific and intellectual level, but at the political muscle level it hasn't delivered yet. And that is what it is going to take to move things like the global warming treaty in this country." Earth Day coordinator Denis Hayes said that "one of the real problems" American environmental groups have in addressing global issues is that their members are not internationalists: "You can't get them as excited about global warming as about Northwest forests." Joshua Reichert of The Pew Charitable Trusts said that U.S. environmental organizations "have to figure out how to work better with colleagues in other countries. They are not good at it yet."

In fact, American environmentalists worked long and hard at the Earth Summit and other international forums to establish common ground and create common agendas and strategies with activists from around the world. To an extent, they were successful. But there appears to be inadequate effort on a continuing basis to build an international environmental coalition. American NGOs can be an important part of that building process, but only if they approach it in the right way. Konrad von Moltke of Dartmouth College, himself an activist, said: "The

U.S. movement is still hung up on notions of leadership. Some groups, like the World Wildlife Fund, have made extraordinary progress, but they are not typical. Most U.S. environmentalists expect to dominate in international settings. It is all very unconscious, but it is not the right approach to this business."

Over the years, American environmentalists have gradually enlarged their vision of what must be done to protect and preserve the global commons and their conception of their role in that process. At least since the Stockholm conference in 1972, for example, some members of the environmental community have focused on linking economic development and reduction of global poverty with the task of protecting global resources and ecosystems. Pressure from U.S. environmentalists and those from other countries helped to reform the ecologically destructive policies of the World Bank and other Bretton Woods financial institutions, although those institutions still fund some ecologically destructive infrastructure projects in developing countries. American environmentalists have increasingly and successfully insisted on participating in diplomatic negotiations about the global environment, and in many cases, as with the global warming treaty, they have been the prime movers behind such efforts.

As with their domestic activities, however, American environmentalists still take an excessively narrow view of what they must do and where they must operate to help preserve the global environment. Brent Blackwelder of Friends of the Earth, itself an international organization, insisted: "The environmental community has to devote more resources to these issues. Now many groups fight narrow battles and score victories that will be overturned by global events. We have to get at the root causes of these issues or we will be continually on the defensive." It is not enough, for example, simply to accept the rising power of the global corporations and try to placate them with market mechanisms such as pollution-trading credits. Certainly those tools should be employed wherever they will accomplish their goals, but they will not by themselves slow the global economic juggernaut or mitigate some of its effects on the global environment. What environmentalists must do, as Kathryn Fuller of the World Wildlife Fund said, is "provide alternatives to the Wild West concept of globalization." Given the unsustainable consumption levels in the United States and the vastly increased consumption sought by the developing countries, "there is no real solution except to fight it out," she said. Environmentalists need to press the U.S. government and, in alliance with NGOs around the world, the governments of other countries, to reform international trading regimes and institutions such as the World Trade Organization so that the concepts of environmental protection and sustainable development are given equal status with the removal of trade barriers. American NGOs have a special responsibility because, as Karl Grossman noted, "the United States is the economic engine of the world and has a major influence on both consumer behavior and the activities of multinational corporations."

Because efforts to safeguard the global environment would be doomed in a world of international political anarchy, American environmentalists need to join in the quest for a workable system of global governance. Such a quest must begin with reform and strengthening of the United Nations—there is too little time to start from scratch. NGOs in the United States could start by joining efforts to have their own government meet its financial obligations to the United Nations and to help turn it into a more democratic institution by abolishing the veto of permanent members of the Security Council. Environmentalists can take the lead in demanding the creation of a comprehensive system of hard international environmental law and a global environmental protection agency with the authority to enforce those laws. An international "green police force," which some have suggested, may not be the answer. The idea of adding another military force to an already overarmed world is a bit chilling. But there needs to be some mechanisms of coercion in addition to market mechanisms, such as pollution credits and taxes, to ensure adherence to international environmental standards. Admiral Sir Julian Oswald, who held the title of First Sea Lord (head of the British navy), suggested that national military forces might be needed to enforce international environmental treaties.[32] Such applications of military power would, however, undoubtedly stir up fierce diplomatic hornets' nests.

Environmentalists need to be engaged in efforts to end armed conflict and the weapons trade, for, as we see over and over again, the environment is inevitably one of the tragic victims of war. International security scholar Thomas F. Homer-Dixon cautioned that global environmental stresses "may come to dominate other factors affecting the international system" in the coming decades and be a major cause of armed conflict.[33]

Finally, if the global environment is to be rescued, it will be only with the broadest possible public participation. Mainstream environmental groups should take on the task of encouraging and assisting grass-roots activism in the global environmental endeavor by offering training as well as technical and material assistance. If national governments are losing their control over events, local and regional governmental and nongovernmental institutions will have to fill the gap. Maurice Strong, who was disappointed in the failure of national governments to follow up on the agreements reached at the Earth Summit, found cause for optimism in the grass-roots response. More than two thousand towns and cities, he noted, have adopted their own versions of Agenda 21.

Many grass-roots activists already recognize and accept their responsibility for engagement in the struggle for the global environment. Richard Moore, director of the Southwest Network for Environmental and Economic Justice, noted that his group does not have the luxury of ignoring international issues. A number of the communities it serves are near the Mexican border and suffer from air and water pollution from the *maquiladoras*—companies that set up manufacturing operations near the border to benefit from lower-wage labor and lax environmental enforcement. One chemical company "that poisoned our people," he said, fled for Mexico, where enforcement of environmental laws is lax, leaving jobless workers behind. "We eliminate poisonous chemicals in the United States, but our brothers and sisters around the world are getting triple exposure. So we have to be part of a world environmental movement, a world justice movement. The only thing that is saving us now are powerful grass-roots organizations going against a Hughes Aircraft Company or an Intel Corporation. We need to build a strong grass-roots movement around the country and throughout the world."

As analysts at the International Institute for Sustainable Development, a research and policy group, concluded, sustainable development begins with "primary environmental care" at the grass roots. "To achieve sustainable development, people must be able to participate in decisions that affect their lives. To provide for this participation requires a democratic political process with effective and accountable institutions at all levels."[34]

The latter part of the 20th century saw a great awakening on the part of the world's peoples about the grave dangers into which modern civilization is thrusting us. Environmental sociologist Riley Dunlap noted: "At the end of the twentieth century globalizing trends in commerce, travel, communication, information and environmental risk have begun to affect popular opinions regarding the environment. Awareness of risks in a changing environment is now appearing in most countries—in less developed economies as well as in industrialized states. . . . But concern among the public at large appears to exceed that of their governments."[35]

The American environmental movement will have to be at the heart of an international campaign to make governments pay due heed—and make appropriate responses.

Chapter 9
Transforming the Future

Rachel Carson dedicated *Silent Spring* to physician, theologian, musician, and humanist Albert Schweitzer, whom she quoted at the beginning of her book as saying: "Man has lost the capacity to foresee and to forestall. He will end by destroying the earth."

But we do have the capacity to foresee. Science and technology have given us increasingly precise tools with which to monitor the effects of human activity on the biosphere and on human health and to predict where those effects are leading us. The modern environmental movement arose out of the foreknowledge that our civilization is moving swiftly along the road that Carson warned would lead ultimately to disaster. If there are still those who decline to acknowledge the darkness that lies ahead, it is not because of any dearth of warning signals.

I also believe that we yet have the capacity to forestall destruction. That is what the environmental movement is about. That is what this book is about. The mission of environmentalism is to mobilize society at all levels to confront the danger and disorder into which human activity has propelled us and to guide us to a safer, saner way of living on this planet, now and in the time of our posterity. Environmentalism has never been about catastrophe. It is about alternatives, about changing course, about transforming the future.

So far, however, environmentalism has been unable to bring us to that other road that, Carson told us, "offers our last, our only chance to reach a destination that assures the preservation of our earth."[1] In substantial part because of the efforts of environmentalists, many of the threats to our bodies and our surroundings created by our industrial, technological, high-consumption, and wasteful civilization have been ameliorated and some wildness has been preserved. But a century after the birth of conservation and thirty years after the first Earth Day, the environmental movement in the United States has yet to engage fully with the fundamental causes of our worsening ecological predicament. As several environmental leaders acknowledged when being interviewed for this book, they are winning battles but losing the war.

The fumes from smokestacks and incinerators, the exhaust from cars and trucks, the chemicals on our food and in our water are only proximate causes of the pollution that fouls our environment, enters our flesh, and threatens the structure of life on earth. Corporate avarice, a Western society obsessed with consumption and possession, and, in the poorer countries, too many humans struggling to survive and prosper on a landscape that cannot meet all their needs are only the immediate agents of the destruction of forests and farmland and fisheries, the loss of space and beauty, the cause of an overstressed ecosphere. Our environmental dilemma is more profound. It springs from the systems, institutions, values, and habits of thought created by humans over recent centuries to manage, sustain, and order our civilization. Or, rather, it springs from a failure of those systems, institutions, values, and habits to change, to adapt, to respond to feedback that signals danger ahead.

In the 21st century, it will be the task and the duty of the environmental movement to address not just the symptoms but also the underlying causes of ecological dysfunction. As Second Nature's Anthony Cortese commented, "We have to think systematically about politics, health, and the economy and how these things relate to the number of people on the planet and people's values. These things cannot be dealt with by thinking about them one at a time."

In recent years, environmentalists have begun to look at ecological problems holistically. They are seeking to protect not just the owl but also the forest and the trees, the salmon, the watershed, and the quality of the forest air. They have yet, however, to adopt a holistic approach to the *practice* of environmentalism. Taking a modest view of their place in society, they seem unable to accept that to succeed they will have to play a major role in many arenas at the same time: They will have to play a role in changing our political and economic systems. They will have to help direct the course of the scientific enterprise as well as make use of its results. Instead of lamenting the shortcomings of education in the United States, they will have to find ways to be instruments of its reform. Instead of raging against the global economy, they will have to help tame it and make it an instrument for fulfillment of human needs instead of social and ecological decline. Environmentalists can and should encourage the movement of churches toward a new theology and offer a moral center for people throughout the United States and the world who are searching for values to guide their lives.

Many, if not most, of the proposals made in this book are well beyond the capacity of the American environmental movement as presently constituted to carry out. Today's environmental organizations—national, regional, local—do not have, by orders of magnitude, the money, personnel, skills, power, standing, or confidence to attempt the fundamental economic, political, and social reforms essential to lead us to a prosperous, sustainable future. To move forward in the new century, the movement will have to become bigger, stronger, richer, more resourceful, more diverse, and more confident of its ability to take a broader role in national and international affairs.

To environmental organizations, some of which are now struggling to survive, such a future may seem beyond reach. But American society has always welcomed change, growth, and regeneration. Only think of how different, how much more central in our society, today's environmental movement is from the nascent conservation movement of a century ago. Think of how, since the first Earth Day only thirty years ago, it has grown and evolved, forced itself on the public consciousness, and compelled the government to act and the corporations to respond.

Given the immensity of the stakes, the continued growth and evolution of the environmental movement is not only possible; it is inevitable. Without a significant change in the way the world currently operates, the coming years will at best be more crowded, hungry, dirty, diseased, hot, ecologically impoverished, and unlovely. Some futurists warn that without a change of course, the 21st century could witness a reversion to barbarism. One scenario suggests that a continuation of business as usual would lead to "cultural disintegration and economic collapse, a degeneration of civilization into a primitive world of all-against-all."[2]

That need not happen. We possess sufficient knowledge as well as the technological and social organizational skills to avoid such catastrophe. What is still lacking, however, is the will to act decisively. Given the erosion of governmental authority to produce change, particularly in the United States in recent years, the nongovernmental sector will have to provide much of the will and impetus for change. The American environmental movement has a leading role to play in such an effort. First, however, it must gird itself.

Transforming the Movement

The most immediate and urgent task facing the environmental movement, therefore, is to build its *capacity* to be an agent of broad social change, to focus on what Mark Van Putten, president of the National Wildlife Federation, called "the tough, long-term, not often satisfying work of institution building."

As the new century progresses, the environmental movement is likely to look very different from the way it looks today. For a while, at least, as political power flows from the federal government to states and local communities, the movement is likely to grow increasingly decentralized. Although it will be necessary to maintain maximum leverage on federal policy, the ability of the national environmental groups to move forward with their agendas will depend on their aggressively mobilizing a grass-roots membership base to serve as political shock troops rather than the passive dues-payers that many are today. With such an aggressive army, the environmental movement will be able to demand more effective laws and more responsive institutions. For the movement to move forward on a wide front, there must be close collaboration and mutual support among national, regional, and local activist organizations—including environmental justice and community-based development groups. Collaboration between the national environmentalists and local activists must be a two-way street, with knowledge, ideas, and leadership moving in both directions.

Citizen organizations fighting local environmental battles, churches, and, wherever and whenever possible, union locals can be a source of strength and inspiration. West Harlem Environmental Action, which has an agenda that includes the environment, economic development, and social justice and addresses all these issues with political action, serves as a model that could be emulated by environmental organizations at all levels.

It may be that to reach a higher level of efficiency and effectiveness, the environmental community will have to go through a period of rationalization. Although diversity is one of the strengths of the movement, cutthroat competitiveness and overlapping functions are not. The early years of the 21st century are likely to see mergers of national organizations with similar missions that will enable them to make the most of their resources. Some groups will, no doubt, drop by the wayside in the process. But it is also certain that new institutions will arise to perform the tasks that existing groups do not or cannot. Some of these groups will eschew tax-exempt status in order to participate directly in the political process, as the League of Conservation Voters and a handful of other organizations do today. Others will decline funding from corporations and wealthy individuals so that they can operate more boldly without fear of losing their economic underpinnings. New institutions will emerge to bridge the gap between those that operate at the national and global levels and the grass-roots activists, whose energy, spirit, and anger, as well as their ground-level knowledge, will fuel the next stage of environmentalism.

Lois Gibbs's Center for Health, Environment and Justice plays such a role today. Working with thousands of local activist organizations around the country, the center provides information and technical assistance to help those groups organize and fight and win their battles in the arena of local politics. But the center operates on only a limited number of issues and with scant resources. There is little cooperation between it and other national environmental groups.

Both old and new environmental institutions will need to be more proficient in acquiring financial resources. Fund-raising from members has its limitations—the Sierra Club, for example, has been deliberately limiting its membership to 0.5 million because expanding the membership would cost more than it would bring in. But there is still an untapped potential donor base among the millions of Americans who profess to be environmentalists.

The foundation world may be an even more promising vein to be mined for economic support in the near future. With the economic boom and the bull market of the 1990s, many of the foundations have seen their endowments swell to unprecedented levels and are looking eagerly for worthy projects. Certainly, investment in an enlarged and strengthened environmental movement is a good investment. But the foundations that support environmental causes need to find a common strategy to replace what today is a largely scattershot approach. There are lessons to be learned from the successes of the right-wing foundations, whose targeted, concentrated strategy have played a major role in transforming the political life of the United States in recent decades. The foundations can play a crucial role in capacity building within the environmental movement. At least some of them, however, will have to change their current policy of declining funding for long-term institutional support.

Acquiring sufficient economic resources to carry out its mission probably will require a new, much more intensive approach by the environmental community. It is possible that in the future, some environmental organizations will also be entrepreneurial enterprises earning substantial amounts of revenue that, instead of being distributed as profit, is invested in the great task of preserving our habitat. This is not as radical a thought as it may appear at first glance. Grass-roots organizations have always tried to raise money with very small capitalist efforts such as bake sales. A number of environmental groups already are engaged in various forms of money-raising endeavors, such as the sale of books, records, T-shirts, and similar items or the packaging of wildlife tours and vacations.

As described in chapter 5, the Natural Resources Defense Council (NRDC), in partnership with a community-based economic development group, has been attempting to build a large-scale industrial enterprise, the Bronx Community Paper Company. This is an innovative and risky project for the group, which has concentrated on litigation and education of the public on issues and science since its founding more than three decades ago. The project has proved difficult, and NRDC has forsworn taking revenues from it. But the effort demonstrates that the thing is not impossible. Should the environmental effort continue to link up with the community-based development movement in this way, the focus and scale of grass-roots fiscal activities would change considerably. Indeed, because it is now recognized that a healthy environment and a sustainable economy go hand in hand, future efforts to improve local economies and social conditions may be seen as integral to efforts to preserve local ecology.

The very definition of what constitutes an environmental organization may become blurred in the coming years. Just as some environmental groups may also be revenue-generating businesses, many businesses that have a stake in a society that protects its environment—for example, organic food producers and retailers, community-supported farms, alternative energy and appliance producers, mass transit companies and manufacturers of alternative fuel vehicles, local businesses that challenge the global economy, and many more—may in coming years be seen as part of the environmental movement.

Expertise in revenue raising is only one of the skills the environmental organizations will need to acquire or upgrade if they are to be an instrument of significant social change in the 21st century. Having relied so heavily on litigation and lobbying to achieve their goals, many of the organizations are now top heavy with lawyers. Allen Hershkowitz, the scientist who has devoted ten years of his life to building the Bronx Community Paper Company, believes that the movement now needs to stock up on economists and individuals with M.B.A. degrees so that they can be players in the marketplace. These experts can help make their organizations run more efficiently and make better use of their financial resources, including investment of their endowments; they can improve analysis trends and help plan programs that integrate care for the environment with care for local economies, jobs, and communities. Such expertise can be invaluable in negotiations with industry and government.

It is also imperative for the national environmental groups to build a cadre of aggressive, competent field organizers if they are to tap and mobilize the broad latent support for their goals that is suggested by the public opinion polls. The National Environmental Trust demonstrated how vital these skills are when it played a central role in plugging the dike against the torrent of retrograde legislation during the 104th Congress. The trust, created with funding from The Pew Charitable Trusts, augmented the national movement where it was weak: in political professionalism, field organizing, and use of the media to achieve its goals.

Building a firm, permanent base of public support nationwide, however, and translating that support into substantive action will require intensive, sustained organizing efforts by a wide spectrum of well-funded environmental organizations across the country.

There already are a substantial number of superb scientists in or closely associated with the environmental movement, but given the central role of science in environmentalism, far more resources are required. To participate in the process of civic science, the process of reviewing science and technology before it is deployed, the environmental community will require substantially increased scientific capacity. It will need to recruit chemists, biologists, toxicologists, and physicists as well as ecologists with the credentials, training, and intellectual firepower to sit as peers at the table with scientists from academia, industry, and government. Because environmentalists play such a vital role as intermediaries between science on one hand and the media, the public, and government on the other, environmental groups at all levels are in special need of staff that can communicate the scientific message clearly and accurately. There are only a handful with such skill in the movement today.

The environmental movement generally needs to become more proficient in getting its message across to the media and the public. Skills in this area have been improving, especially since the formation of Environmental Media Services in the 1990s. Run by veteran journalist, environmentalist, and political professional Arlie Schardt, the organization conducts concentrated media campaigns on environmental issues. It played a key role, for example, in blocking plans by the Disney organization for a historical theme park in northern Virginia. It also conducts regular briefings for reporters in major media centers. But virtually all of the environmental organizations need to invest more in media operations if they are to do a better job of mobilizing the public and increasing their influence on policy decisions.

One problem they will have to address, however, is the declining will and ability of the traditional media to transmit information and ideas that enable Americans to participate knowledgeably in the democratic process. As the media, with some honorable exceptions, become increasingly concentrated into conglomerates concerned almost exclusively with the bottom line, more and more emphasis is placed on entertainment and less on news. Indeed, the line between the two is thinning, and the end product is what some have dubbed "infotainment." Instead of providing the intellectual nourishment for an informed electorate that so impressed Alexis de Tocqueville in the early years of the Republic, the media in our time are becoming the opiate of the people.

Fortunately, in the last years of the 20th century, a powerful alternative means of public communication emerged—the Internet. As people in the United States and around the world increasingly become connected to the Internet and the World Wide Web, environmentalists will be able to bypass the corporate media and get their message out on an electronic "green samizdat." The Internet can be a transforming populist instrument, but only if it is used effectively. The environmental movement as a whole will need to develop a strategy for using the Internet and pool resources to make it yield the biggest bang possible. For example, there could be a database describing how local environmental groups around the country have dealt with such problems as hazardous waste dumps, polluting industrial sites, brownfields, and other land use issues. With such information available to them, communities facing the same issues would not have to reinvent the wheel. The community-based development groups have had such databases in place for years. Organizations need not give up their own Web pages, but there will have to be a central site for the environmental messages that will reach the largest possible audience with the greatest possible effect.

Fred Krupp of the Environmental Defense Fund said that he anticipates a "fourth wave" of environmentalism, which "will see communities solving environmental problems using the power of harnessed information. It will weave together the grass roots and the nationals through technology. There will be more bottom-up empowerment and a tremendous trend toward pluralism in the movement because there has to be if we are to solve our problems."

A number of those interviewed for this book expressed the view that the environmental movement requires new, younger leadership because, they said, some of those who run the major groups have grown stale over the years. I do not entirely agree—perhaps because I recently received my Medicare card in the mail. First, the leadership of most of the national groups has already turned over, in some cases several times, since the first Earth Day. Second, although the movement can always use fresh, aggressive, energetic younger leaders with new ideas, it also needs the wisdom, experience, and patience of those who have been around the track a few times. People such as John Adams of the Natural Resources Defense Council and Gaylord Nelson of The Wilderness Society bring priceless continuity and stability to the movement. Environmentalism would be the beneficiary if David Brower, its "archdruid," who is now in his eighties, turned out to be immortal. The environmental cause suffered a substantial loss in 1999 when Michael McCloskey, chairman of the Sierra Club and an activist for forty years, retired. New, able young leaders undoubtedly will pick up the banners, but the environmental movement needs the steadiness, the ballast, that its veterans can provide as it moves into the new century. It is of more than symbolic significance that Denis Hayes, a key organizer of the first Earth Day, is playing a major role in organizing the first Earth Day of the new millennium.

Transforming Society

It has been said that the environmental movement is on the cutting edge of a new postindustrial, postmodern society that will be markedly different from our current industrial society with its attendant economic inequities, social injustice, and degradation of the biosphere. As the new century begins, however, environmentalism has barely scratched the skin of the old order. The old pattern of mass production and mass consumption, of polluting technologies, misuse of land, and waste of resources, is still in place and is being reinforced by globalization and the politics of laissez-faire conservatism.

Transformation of the economy is an essential task, a task that will be especially difficult in the United States as long as the late-20th-century flood of prosperity continues. When the good times roll, nobody wants to hear that they are rolling downhill toward an ecological blank wall.

Environmentalists will have to make a persuasive case that there can be a different kind of prosperity within a different kind of economy, an economy that provides the necessities and comforts of life but places less value on ever-expanding consumption and greater value on aesthetics, on spirituality, and on learning and leisure as well as on the health of the natural world and the welfare and happiness of its inhabitants. It will husband energy and material resources and reuse them in a circular economy rather than throwing them away as pollution of the ground, air, and water after a short lifetime of linear use. It will be an economy in which the planet's finite resources are shared more equitably so that no member of the human race goes hungry or lives a life of extreme poverty.

To achieve such an economy, the environmental movement will be required to do more than exercise its powers of persuasion—it will have to participate in its creation. One of the ways it can do so is to enlist the powerful machinery of the market in the service of economic transformation—to create a capitalism with a green face. As labor organizer Dan Swinney said, "We don't have to worship the free market. We can use its strengths and jettison what is bad."

As Swinney and others have suggested, one powerful way in which environmentalists can use the market to achieve their goals is to take over capital formation and the means of production where and when they can—to become green capitalists. Allen Hershkowitz suggested that environmentalists "infiltrate" Merrill Lynch, Goldman Sachs, and other investment banking and financial institutions. Stephen Viederman said that he wants shareholders to exercise their right to demand more responsible corporate policies, and a number of commentators suggested more emphasis on investment in companies with records of environmental and social responsibility. The movement is already pressing for the use of market tools such as pollution-trading permits and the use of tax policy as a incentive to discourage pollution and waste and encourage capital formation, job creation, and a decrease in adverse environmental effects. The idea of shifting taxes from wages and earnings to pollution and resource depletion—without increasing the total tax burden—is gaining currency among mainstream economists and is being considered seriously by environmentalists and even some business and labor leaders. As a monograph of the policy organization Redefining Progress stated, "By shifting some of the focus of taxation from 'goods' like work to 'bads' like pollution, the tax system would validate the public's sense that work and saving are good and that wasteful consumption is not."[3]

Adaptive management in the marketplace is an efficient way to test approaches to reducing the human effect on the environment and to do so at the lowest possible cost. Environmentalists also should insist that economic policies be based on real measures of social and physical welfare rather than on the dubious precision and utility of the gross national product. Use of market tools must be backed up by a strong federal and state regulatory regime that prevents backsliding and encourages innovation to meet market goals if venal interests are not to quickly crush the public interest.

Assuming that the environmental community musters the will and the resources to help alter the trajectory of an economy now in ecological free fall, it will have to be prepared to come up with a realistic, carefully thought out strategy for making the transition from the current flawed system. Yale University's Daniel C. Esty insisted that "half the battle will be won or lost with transition policies" and added that "the environmentalists have not paid much attention to the transition—they are too impatient and too moral." The imperative of solving our ecological problems does not mean we should rush forward with economic reforms that cost workers their livelihood or communities their stability. We cannot have ecological and economic sustainability without social sustainability. Any transition policy has to be protective of jobs. Such a policy must also avoid major shocks to the economy and be adopted in ways that ensure public acceptance. For example, Esty proposed a substantial increase in the gasoline tax to provide a disincentive to individual automobile use, but he recommended a tax that is phased in over twenty-five years or more and that increases by only a few cents per year.

Ecotrust, formed in the early 1990s to integrate conservation and economic development in the Pacific Northwest, is a model for a new approach to making the transition from an unsustainable industrial economy. Working with local communities, Ecotrust has helped to plan and find capital for a restructured economy based on conservation and careful use of local and regional resources such as salmon. At the heart of the Ecotrust approach is the understanding that a healthy natural community now depends on a healthy human community.

If the economy is the engine of social change, the driver of our ecological future, politics is the mechanism that steers the machine. Regarding the environment, at least, we have been steered well off course in recent years. As political scientist Michael Kraft commented, our environmental policies are badly in need of revitalization, but there is unlikely to be reform without a major change in the balance of political power.[4] However, the dynamics of the political system as it currently functions in the United States tend to preserve the status quo rather than to create opportunities for altering the balance of power. Political scientist John Rensenbrink, a leader of the Maine Green Party, wrote in 1999: "The links and channels that are supposed to connect people to their government and to each other, and to connect their government to them, are barely functioning. Politicians take their signals from the entrenched moneyed interests." The major political parties, he wrote, "are more willing to serve the special interests than they are to provide real links between the government and the people."[5] He concluded that the current political system will not be able to address our ecological problems, and therefore "we need a new constellation of power."[6]

The American environmental movement can—and, if it is to move forward in the coming years, must—participate actively in changing the constellation of political power. The rise of Green parties in Europe demonstrates that environmentalism can change the face of politics. Although the Greens are unlikely to take national power in the United States in the absence of proportional representation, here, too, environmental politics has the potential of forcing badly needed changes in the political and policy-making processes. As a broadly based social movement with wide popular support, the American environmental movement has latent political strength that it has barely tested so far. Some scholars think that the transcendent importance of environmental problems will inevitably bring them to the top of our future political agenda.

To change the nature of American politics, the environmental community has to build its capacity to engage in and influence the electoral process. It will need to acquire more financial resources and trained personnel to throw into the political arena, including a cadre of professional field organizers. Older institutions such as the League of Conservation Voters will need to be strengthened, but undoubtedly new organizations will emerge that are free of tax constraints on their political activities. Because money makes an unequal battle out of politics for environmentalists and other public interest activists, the community needs to throw its full weight behind the drive to reform the campaign finance system. It needs to challenge the legal principles that corporations are persons entitled to the same free speech as individual Americans and that money is equivalent to speech.

Perhaps most important, the environmentalists will need to find allies if they are to alter the current trajectory of American politics. They need finally to recognize and acknowledge that the decline of the environment is *not* an issue distinct from other flaws in our society, and they need to realize they are in the same boat with other groups of Americans who are thwarted and victimized by the status quo. Once they make that leap, they can join forces with them in a broad-based political coalition.

With more resources, new institutions, and new allies, the environmental community ought to be able to establish a broad base of political power by contesting elections from the village level to Washington, D.C. Green parties can win office at the state and local levels. Candidates committed to environmental goals and supported by environmentally astute political organizers can win at all levels. Because informed citizens with a grasp of the civic process are vital to democratic politics, high-quality school systems and curricula are essential. It is critically important, therefore, that environmentalists and their allies win seats on boards of education. The recent successes of right-wing fundamentalists in taking control of such boards is a threat to the environmental enterprise as well as to democracy itself.

Because the environmental enterprise is now global in scope, American environmentalists will have to become increasingly proficient at geopolitics as well as domestic politics—although the two are intimately connected. Environmentalists can join with human rights organizations, peace groups, and those seeking economic justice for developing countries to press for a reformed and strengthened United Nations as well as reformed Bretton Woods institutions and other institutions of international governance. They must demand that such institutions create a system for controlling global corporations and requiring the corporations to meet real human and ecological needs. They need to insist on the development of a body of hard international law—binding commitments subject to penalties such as sanctions if violated—to protect the environment. Among other things, international law should provide for those who despoil the ecosphere to be punished for crimes against humanity.

Environmentalists can lobby for substantially increased resources for the Global Environment Facility, a multilateral fund to help less-developed countries meet global environmental standards, to help the poorer countries prosper in ways that preserve rather than degrade the global environment. They can put increasing pressure on government in the United States and around the world to require that the World Trade Organization and the North American Free Trade Agreement give adequate regard to the environment and workers' rights in their onrushing drive toward freer trade and economic integration. The American movement can also do much more to help create a worldwide alliance of environmental activists. But it must do so without arrogance, by learning as well as teaching and by following as well as leading.

Finally, if environmentalists are to protect the global environment, they must join in the effort to stop people from killing one another. Although slow-motion assaults on the ecosphere from global warming, acid rain, toxic materials, loss of arable land, inadequate freshwater, and the extirpation of species are the gravest threats to life on earth, modern warfare causes traumatic, often irremediable injury to the living planet and its inhabitants. During the 1999 fighting in Kosovo, Greens throughout western Europe protested the "ecocide" the heavy NATO bombing was causing, and their protests did not go unheeded by governments. Environmental groups in the United States, however, were for the most part silent on the war.

This discussion of what the environmental community must do to move forward in the 21st century has focused thus far on concrete matters such as economics, politics, science, and social change. But the real strength of the environmental movement, the source of its actual and latent power, comes from its values, its broad perspective on what is important in our lives and how we should live. Those values have changed little over the years: Care for nature. Respect for life and awareness of its fragility. Awe and humility before the Creation. Love of place. An acceptance that we are responsible for the future as well as the present. An understanding that our health and happiness depend on protection of our habitat, both in our immediate vicinity and throughout the planet; that we break any link in the chain that holds life together at our own peril.

By the end of the 20th century, many Americans appeared to have taken those values as their own. But they seemed to have done so in a shallow and possibly transitory way. To acquire the resources and power they will need to succeed in their labors in the new century, environmentalists will have to make sure that these values deepen and become central in the national consciousness. Comments from many of those I interviewed for this book, both within and without the movement, however, suggest that environmentalists at all levels, preoccupied by fighting the intense battles of the moment, are forgetting, or are neglecting or failing to articulate, those values. Patrick Parenteau said that he sees the movement "coming up with strategies for accommodation, not strategies based on the values we started out with. Maybe people are worn out." Charles Little expressed concern that "current environmentalists are not approaching the great ecological crisis with outrage, alarm, and a clear sense of what is right and wrong." David Hahn-Baker said that he sees "real confusion in our society about issues of value and issues of worth—the desire to count things in dollars and cents overwhelms our ability to count things in terms of goodness and what gives meaning to our lives. The environmental community has to find a way to reorient people to what things we should value. It is not fully engaged on that battlefield." Tim Hermach insisted that "the environmental movement has to figure out what it stands for and then stand for it."

Veteran environmentalist Philip Hocker stated that it will take a "moral epiphany" within American society if we are to do what is necessary to save the world. A fundamental mission of environmentalists in the coming years will be to set the example and evangelize for that epiphany.

Transforming the Future

Judeo–Christian culture's first story tells of a man and a woman who lived in a verdant garden in harmony with all other life, subsisting without toil on the abundance provided by the Creator. But they were cast out after taking more from the garden than was permitted and gaining forbidden knowledge and were required thenceforward to labor for their sustenance.

As it turned out, however, the place of exile into which the man and woman and their progeny were thrust was also a beautiful and bountiful garden. Although they would have to earn their bread by the sweat of their brow, if their descendants tended the garden well, it would provide them with all they required in endless variety. When cultivated with wisdom and humility, the land flowed with milk and honey.

In our time, it seems, we are on the verge of losing this other garden, of once again being cast out of Eden. We have forgotten the lessons of Genesis. We are taking more from the garden than is permitted. The knowledge we have gained is being used to degrade and destroy the Creator's bounty. We have become too fruitful and have multiplied too much, and we are failing to replenish the earth. Once again, human folly and frailty have placed us in danger of being thrust into exile, this time not by an archangel with a flaming sword but by the consequences of our own science and technology, by our inadequate economic and political arrangements and the failure of our institutions, by our own excessive numbers, by our ignorance, greed, and aggression and our inability to correct injustices inflicted on our fellow humans and other creatures with whom we share the garden.

There is still time to save the garden and to keep us from exile. Having, through our own works, reached a new stage of coevolution with the earth, we must from now on actively manage the biosphere instead of simply use it. We will have to accept responsibility for the welfare of all life and for the viability of the systems that support and encourage life. We will have to stop taking more than our share of what the garden can provide—stop managing resources for maximum utility and instead manage to preserve them from exhaustion and degradation and to ensure that they are available for our posterity. Many of us will have to rethink our understanding of what life is about and change the way we live our lives.

We already know all this. And we possess sufficient knowledge and tools with which to transform the future. Our science and technology have the capacity to restore much of what has been harmed; we have the ability to organize ourselves politically for action from the local to the global level; our economies still generate enough wealth to meet the needs of the transition; our ability to communicate information and ideas to one another is growing with exponential speed; our churches are finally awakening to their obligation to help preserve Creation.

And we have environmentalism to warn us where our careless gardening is leading and to show us how and where we must do a better job of managing for the future. Now the environmental movement will have to take on a wider role. It will have to take a substantial part in creating a society that has the capacity and the will to end the ecological folly and convey us safely through the new century.

The rise of environmentalism has let us glimpse the possibility of a brighter, more rational and secure future. In the coming years, environmentalists face the enormously difficult and daunting challenge of translating that vision into actuality. They need not be dismayed at the prospect. "If you have built castles in the air, your work need not be lost; that is where they should be," wrote Henry David Thoreau at the conclusion of *Walden*.

But then he added: "Now put the foundations under them."

Notes

Unless otherwise indicated, all direct and indirect quotations in the main text are from interviews by the author.

Chapter 1. The Story until Now

1. George Perkins Marsh, *Man and Nature,* edited by David Lowenthal (1864; reprint, Cambridge, Mass.: Harvard University Press, Belknap Press, 1965).
2. For a fuller account of the origins and development of the American environmental movement, see, among other sources, Philip Shabecoff, *A Fierce Green Fire* (New York: Hill & Wang, 1993).
3. Robert Gottlieb, *Forcing the Spring: The Transformation of the American Environmental Movement* (Washington, D.C.: Island Press, 1993), pp. 10–11 ff. Gottlieb maintains that these efforts were, in fact, central to environmentalism.
4. Samuel B. Hays, *Beauty, Health, and Permanence* (Cambridge, England: Cambridge University Press, 1987), p. 4.
5. Aldo Leopold, *A Sand County Almanac* (New York: Oxford University Press, 1948), p. 215.
6. Rachel Carson, *Silent Spring,* 25th anniversary ed. (Boston: Houghton Mifflin Company, 1987).
7. Quoted in *Congressional Quarterly,* 30 October 1970, p. 2728.
8. Mark Dowie, *Losing Ground* (Cambridge, Mass.: MIT Press, 1995), p. 27.
9. Michael E. Kraft, *Environmental Policy and Politics* (New York: HarperCollins, 1996), p. 74.

Chapter 2. At the Turn of the Millennium

1. Jane Lubchenco, "Entering the Century of the Environment: A New Social Contract for Science," president's address at annual meeting of the American Association for the Advancement of Science, Corvallis,

Oregon, 15 February 1997 (text reprinted in *Science,* 23 January 1998; available on-line at http://www.sciencemag.org/cgi/content/full/279/5350/491).

2. Peter Raven, closing address to the Nature and Human Society Biodiversity Forum, Washington, D.C., 30 October 1997.
3. Cheryl Simon Silver with Ruth S. DeFries, *One Earth, One Future: Our Changing Global Environment* (Washington, D.C.: National Academy Press, 1990), p. 2.
4. Joel E. Cohen, *How Many Humans Can the Earth Support?* (New York: W. W. Norton & Company, 1995), app. 2, p. 400.
5. Robert Engelman, *Why Population Matters* (Washington, D.C.: Population Action International, 1997), p. 13.
6. United Nations Secretariat, Department of Economic and Social Affairs, Population Division, *World Population Projections to 2150* (New York: United Nations Secretariat, 1 February 1998), p. 1; available on-line at http://www.undp.org/popin/wdtrends/execsum.htm (24 August 1999).
7. Lester R. Brown and Jennifer Mitchell, "Building a New Economy," in *State of the World 1998,* edited by Lester R. Brown (New York: W. W. Norton & Company, 1998), p. 168.
8. Raven, address to Nature and Human Society Biodiversity Forum.
9. Benjamin R. Barber, *Jihad vs. McWorld* (New York: Ballantine Books, 1995), p. 54.
10. Philip Shabecoff, "After Decades of Deception, a Time to Act," in *An Appalachian Tragedy: Air Pollution and Tree Death in the Highland Forest of Eastern North America,* edited by Harvard Ayers, Jenny Hager, and Charles E. Little (San Francisco: Sierra Club Books, 1998), pp. 188, 193.
11. Environmental Defense Fund, *Environmental Defense Fund Strategic Plan* (New York: Environmental Defense Fund, 1997), pp. 1–2.
12. Lester R. Brown, Michael Renner, and Christopher Flavin, *Vital Signs 1997* (New York: W. W. Norton & Company, 1997), p. 97.
13. John Tuxill and Chris Bright, "Losing Strands in the Web of Life," in Brown, *State of the World 1998,* p. 41.
14. Philip Shabecoff, *A Fierce Green Fire* (New York: Hill & Wang, 1993), p. 183.
15. George Schaller, speech given at conference on Nature and Human Society, Library of Congress, Washington, D.C., 28 October 1997.
16. Anne Platt McGinn, "Promoting Sustainable Fisheries," in Brown, *State of the World 1998,* p. 60.
17. Henry W. Kendall and David Pimentel, "Constraints on the Expan-

sion of the Global Food Supply," *Ambio* 23, no. 3 (May 1994): 204–205.

18. "Water: World Market Would Ensure Supply,"*Greenwire,* 23 March 1998; available on-line to subscribers at http://www.nationaljournal.com/aboutgreenwire.htm.
19. David Korten, *Getting to the Twenty-First Century* (West Hartford, Conn.: Kumarian Press, 1990), p. 1.
20. Ibid., p. 11.
21. Enterprise for the Environment, *The Environmental Protection System in Transition* (Washington, D.C.: CSIS Press, 1997), p. 2.
22. Barry Commoner, "What Is Yet to Be Done," *New Solutions* 8, no. 1 (1998): 82.
23. For a fuller discussion of the Reagan counterrevolution, see Shabecoff, *A Fierce Green Fire,* pp. 203–230.
24. Riley E. Dunlap, "The Evolution of Environmental Sociology: A Brief History and Assessment of the American Experience," in *The International Handbook of Environmental Sociology,* edited by Michael Redclift and Graham Woodgate (Northampton, Mass.: Edward Elgar Publishing, 1997), p. 27.
25. For a fuller discussion of environmental geopolitics, see Philip Shabecoff, *A New Name for Peace: International Environmentalism, Sustainable Development, and Democracy* (Hanover, N.H.: University Press of New England, 1996), pp. 112–127.
26. Walter A. Rosenbaum, *Environmental Politics and Policy,* 3d ed. (Washington, D.C.: CQ Press, 1995), p. 7.
27. Michael E. Kraft, *Environmental Policy and Politics* (New York: HarperCollins, 1996), p. 133.
28. William Ophuls and A. Stephen Boyan Jr., *Ecology and the Politics of Scarcity Revisited: The Unraveling of the American Dream* (New York: W. H. Freeman & Company, 1992), pp. 309–310.
29. Thomas Berry, speech given at conference on Christianity and ecology organized by the Harvard University Center for the Study of World Religions, American Academy of Arts and Sciences, Cambridge, Massachusetts, 16 April 1998.

Chapter 3. Shades of Green: The State of the Movement

1. Angela G. Mertig and Riley E. Dunlap, "Public Approval of Environmental Protection and Other New Social Movement Goals in Western Europe and the United States," *International Journal of Public Opinion Research* 7, no. 2 (1995): 145.

2. Walter A. Rosenbaum, *Environmental Politics and Policy,* 3d ed. (Washington, D.C.: CQ Press, 1995), p. 22.
3. Lynton Keith Caldwell, *Between Two Worlds: Science, the Environmental Movement, and Policy Choice* (Cambridge, England: Cambridge University Press, 1992), p. 86.
4. Bill Devall, "Deep Ecology and Radical Environmentalism" (paper), 1990, p. 15. This book, incidentally, does not discuss deep ecology at any length. The insistence of the deep ecologists that there must be a nature-oriented rather than a human-oriented view of the world does make a substantial contribution to the intellectual and moral underpinnings of contemporary environmentalism and has motivated intense activism in groups such as Earth First! In the workaday, pragmatic context of American society, deep ecology, which sounds to many like a longing for the primitive, is unlikely to be a decisive force in making the changes in our culture and institutions required for a shift toward a more secure ecological future.
5. Adam Werbach, *Act Now, Apologize Later* (New York: HarperCollins, 1997), pp. 68–69.
6. Mertig and Dunlap, "Public Approval of Environmental Protection."
7. Estimates are from Rosenbaum, *Environmental Politics and Policy,* p. 27, and Jonathan Adler, *Environmentalism at the Crossroads: Green Activism in America* (Washington, D.C.: Capital Research Center, 1995), pp. 147–232.
8. Interview with Lois Gibbs.
9. Peter Borrelli, ed., *Crossroads: Environmental Priorities for the Future* (Washington, D.C.: Island Press, 1988), p. 10.
10. Christopher J. Bosso, "Seizing Back the Day: The Challenge to Environmental Activism in the 1990s," in *Environmental Policy in the 1990s,* edited by Norman J. Vig and Michael E. Kraft (Washington, D.C.: CQ Press, 1997), p. 53.
11. In the interest of full disclosure, it should be noted that the author is a board member of Environmental Media Services. It should also be noted that The Pew Charitable Trusts and Rockefeller Financial Services provided substantial funding to support the preparation of this book.
12. Adler, *Environmentalism at the Crossroads,* p. 87.
13. Scott Allen, "Environmental Donors Set Tone," *Boston Globe,* 20 October 1997, p. 1.
14. People for the American Way, "Buying a Movement: Right-Wing Foundations and American Politics," pp. 1–6; available on-line at http://www.pfaw.org/issues/right/rw/ (24 August 1999).

15. Environmental Defense Fund, *Environmental Defense Fund Strategic Plan* (New York: Environmental Defense Fund, 1997), pp. 2–4.
16. Bosso, "Seizing Back the Day," pp. 62–63.
17. Michael E. Kraft, *Environmental Policy and Politics* (New York: HarperCollins, 1996), p. 76.
18. Tom Hayden, *The Lost Gospel of the Earth: A Call for Renewing Nature, Spirit, and Politics* (San Francisco: Sierra Club Books, 1996).
19. Werbach, *Act Now, Apologize Later,* p. 303.
20. David Korten, *Getting to the Twenty First Century* (West Hartford, Conn.: Kumarian Press, 1990), p. xiii.

Chapter 4. Environment, Community, and Society

1. *New Bearings,* The Strategic Vision of Ecotrust, 1993, (Portland, Ore.: Ecotrust) p. 6.
2. Rose Augustine, "From the Barrio to the World," in *Ten Years of Triumph* (Arlington, Va.: Citizens Clearinghouse for Hazardous Waste, 1993), pp. 27–28. (The Citizens Clearinghouse for Hazardous Waste is now the Center for Health, Environment and Justice.)
3. Dana Lee Jackson, "Women and the Ecological Era," in *People, Land, and Community,* edited by Hildegarde Hannum (New Haven, Conn.: Yale University Press, 1997), p. 40.
4. Robert N. Bellah et al., *Habits of the Heart* (New York: Harper & Row, 1985), pp. 284–285.
5. Michael McCloskey, "Social Issues and the Environment," discussion paper for a retreat of the Sierra Club's board of directors, 9 July 1993, pp. 6–12.
6. Brian Tokar, *Earth for Sale* (Boston: South End Press, 1997), p. 117.
7. Estimates were provided by Alice Shabecoff, founder of the Community Information Exchange, which maintains a database on the activities of community development organizations nationwide.
8. Alice Shabecoff, "Green Communities, Green Jobs," in *Strategy Alert* (quarterly publication of the Community Information Exchange), no. 50 (winter 1998): 1–2. In the interest of full disclosure, it should be noted that Alice Shabecoff is my wife.
9. Ibid., p. 2.
10. David Hahn-Baker, "Funding Sustainable Communities," Working Paper, National Network of Grantmakers, Atlanta, Ga., October 1994, p. x.
11. President's Council on Sustainable Development, *Sustainable America: A New Consensus for Prosperity, Opportunity, and a Healthy*

Environment for the Future (Washington, D.C.: President's Council on Sustainable Development, 1996), p. 87.

12. David Korten, *Getting to the Twenty-First Century* (West Hartford, Conn.: Kumarian Press, 1990), p. 97.
13. Deeohn Ferris, "A Call for Justice and Equal Environmental Protection," in *Unequal Protection,* edited by Robert D. Bullard (San Francisco: Sierra Club Books, 1994), p. 298.
14. "Toxic Waste and Race in the United States: A National Report on the Racial and Socioeconomic Characteristics of Communities with Hazardous Waste Sites (Commission for Racial Justice, United Church of Christ, 1987). For more information, see Philip Shabecoff, *A Fierce Green Fire* (New York: Hill & Wang, 1993), p. 241.
15. Mark Dowie, *Losing Ground* (Cambridge, Mass.: MIT Press, 1995), p. 151.
16. Bullard, *Unequal Protection,* p. xvii.
17. Ibid., p. viii.
18. Ernesto Cortes Jr., "Reweaving the Fabric," in *Interwoven Destinies: Cities and the Nation,* edited by Henry G. Cisneros (New York: W. W. Norton & Company, 1993), p. 318.
19. Edward Skloot, "Letter from the Executive Director," in *Annual Report of the Surdna Foundation,* October 1998, p. 7; available on-line at http://www.surdna.org/surdna/edlet.html (24 August 1999).
20. Julie Light, "The Education Industry: The Corporate Takeover of Public Schools," editorial, *Corporate Watch,* 8 July 1998; available on-line at http://www.corpwatch.org/trac/feature/education/index.html (31 August 1999).
21. Bruce Selcraig, "Reading, 'Riting, and Ravaging," *Sierra* (May–June 1998): 61.
22. David W. Orr, *Earth in Mind: On Education, Environment, and the Human Prospect* (Washington, D.C.: Island Press, 1994), p. 2.
23. David W. Orr, *Ecological Literacy* (Albany: State University of New York Press, 1992), p. 108.
24. Lester W. Milbrath, *Envisioning a Sustainable Society: Learning Our Way Out* (Albany: State University of New York Press, 1989).
25. "News Junkies, News Critics: How Americans Use the News and What They Think about It," Newseum Survey by the Roper Center for Public Opinion Research, February 1997. The Freedom Forum World Center, Arlington, Va.
26. Lynne White Jr., "The Historical Roots of Our Ecological Crisis," *Science* 155 (10 March 1967): 1203–1207.

27. Mary Evelyn Tucker and John Grim, series foreword to conference on Christianity and ecology organized by the Harvard University Center for the Study of World Religions, American Academy of Arts and Sciences, Cambridge, Massachusetts, 16 April 1998.
28. Thomas Berry, "The Challenge of Our Times," *Earth Ethics* (fall–winter 1997–1998): 29.
29. Tom Hayden, *The Lost Gospel of the Earth: A Call for Renewing Nature, Spirit, and Politics* (San Francisco: Sierra Club Books, 1996), p. 12.
30. Teresa Watanabe, "The Green Movement Is Getting Religion," *Los Angeles Times,* 25 December 1998, p. 1.
31. E. F. Schumacher, *Small Is Beautiful* (New York: Harper & Row, 1973).
32. Daniel Kemmis, *Community and the Politics of Place* (Norman: University of Oklahoma Press, 1990), pp. 6–7.
33. "The Quincy 'Compromise,'" editorial, *San Francisco Chronicle,* 4 May 1998, p. A22.
34. "The Quincy Library Bill Controversy: Round Table or Square?" *A CLEAR View* (publication of the Environmental Working Group's Clearinghouse on Environmental Advocacy and Research) 4, no. 15 (20 November 1997): 3.
35. Tom Kenworthy, "Negotiating West's Great Divide," *Washington Post,* 6 December 1998, p. A29.
36. Bellah et al., *Habits of the Heart,* p. 296.

Chapter 5. The Business of America: Environmentalism and the Economy

1. Daniel Bell, *The End of Ideology: On the Exhaustion of Political Ideas in the Fifties* (Cambridge, Mass.: Harvard University Press, 1988).
2. John Gray, "The Best-Laid Plans" (review of *Seeing Like a State* by James C. Scott), *New York Times Sunday Book Review,* 19 April 1998, p. 36.
3. "The Bridge to the High Road, Part 2," *Rachel's Environment & Health Weekly,* no. 619 (8 October 1998); available on-line at http://www.rachel.org/bulletin/index.cfm?St=3 (31 August 1999).
4. Lester C. Thurow, "The Boom That Wasn't," *New York Times,* 18 January 1999, op-ed page.
5. Frances Moore Lappé, "Toward a Politics of Hope: Lessons from a

Hungry World," in *People, Land, and Community,* edited by Hildegarde Hannum (New Haven, Conn.: Yale University Press, 1997), pp. 69–85.

6. Jonathan Adler, *Environmentalism at the Crossroads: Green Activism in America* (Washington, D.C.: Capital Research Center, 1995), p. 139.
7. A. C. Pigou, *Economics of Welfare* (1920; reprint, London: Macmillan, 1952).
8. Herman E. Daly and Kenneth N. Townsend, *Valuing the Earth* (Cambridge, Mass.: MIT Press, 1993), pp. 2–3.
9. Hazel Henderson, *Building a Win-Win World* (San Francisco: Berrett-Koehler Publishers, 1996), p. 36.
10. Brian Tokar, *Earth for Sale* (Boston: South End Press, 1997), p. 197.
11. Robert Costanza et al., "The Value of the World's Ecosystem Services and Natural Capital," *Nature* 387 (7 December 1997): 253–260; available on-line to subscribers at http://www.nature.com/.
12. Robert Kuttner, "The Folly of Free Market Worship," *Boston Globe,* 6 September 1998, p. A9.
13. "LILCO to Stop Selling Emissions Credits," *Greenwire,* 4 April 1998; available on-line to subscribers at http://www.nationaljournal.com/aboutgreenwire.htm.
14. National Academy of Public Administration, *Resolving the Paradox of Environmental Protection: An Agenda for Congress, EPA, and the States* (Washington, D.C.: National Academy of Public Administration, 1997), p. 70.
15. Walter A. Rosenbaum, *Environmental Politics and Policy,* 3d ed. (Washington, D.C.: CQ Press, 1995), p. 23.
16. Ibid., p. 12.
17. Michael E. Kraft, *Environmental Policy and Politics* (New York: HarperCollins, 1996), pp. 170–171.
18. Richard D. Morgenstern, William A. Pizer, and Jhih-Shyang Shih, *Are We Overstating the Real Economic Costs of Environmental Protection?* Resources for the Future Discussion Paper no. 97-36-REV (Washington, D.C.: Resources for the Future, June 1997).
19. Richard D. Morgenstern, William A. Pizer, and Jhih-Shyang Shih, *Jobs versus the Environment: Is There a Trade-off?* Resources for the Future Discussion Paper no. 9901 (Washington, D.C.: Resources for the Future, October 1998), p. 14.
20. "Falling Off a Log: RMI Helps Timber-Dependent Towns Diversify," *Rocky Mountain Institute Newsletter* (spring 1998): 8.

21. Kenneth Arrow et al., "Economic Growth, Carrying Capacity, and the Environment," Second Asko Meeting and Statement, 31 August–2 September 1994, unpaginated. Also in *Science* 268 (28 April 1995): 520–521.
22. World Commission on Environment and Development, *Our Common Future* (New York: Oxford University Press, 1987), p. 43.
23. Herman E. Daly, "Sustainable Growth: An Impossibility Theorem," in Daly and Townsend, *Valuing the Earth*, p. 267.
24. E. J. Dionne Jr., "'Smart Growth' Politics," *Washington Post*, 15 January 1999, p. A29.
25. Richard Grossman, "Corporations Must Not Supplant 'We the People,'" *Maine Sunday Telegram*, 4 February 1996, op-ed page. (Grossman is codirector of the Program on Corporations, Law & Democracy.)
26. Paul Hawken, *The Ecology of Commerce* (New York: HarperCollins, 1993), p. 92.
27. Richard L. Grossman and Frank T. Adams, *Taking Care of Business: Citizenship and the Charter of Incorporation* (Cambridge, Mass.: Charter, Ink, 1993), p. 20.
28. Stephen Viederman, "From Prudent Man to Prudent Person: Sustainability and Institutional Investments for the 21st Century" (notes from presentation at Harvard Seminar on Environmental Values, Cambridge, Massachusetts, 12 December 1996), p. 4.
29. William Ophuls and A. Stephen Boyan Jr., *Ecology and the Politics of Scarcity Revisited: The Unraveling of the American Dream* (New York: W. H. Freeman & Company, 1992), p. 311.
30. Victor Papanek, *The Green Imperative* (New York: Thames & Hudson, 1995), p. 46.
31. "The Natural Step Takes a First Step in the U.S.," *Environmental Building News* 5, no. 2 (March–April 1996); available on-line at http://www.ebuild.com/Archives/Other_Copy/Naturalstep.html (31 August 1999).
32. "Monsanto's Enemies Score Direct Hits," *Corporate Watch*, 28 October 1996; available on-line at http://www.corpwatch.org/trac/corner/worldnews/other/238.html (31 August 1999).
33. Claudia H. Deutsch, "For Wall Street, Increasing Evidence That Green Begets Green," *New York Times*, 19 July 1998, Sunday business section, p. 7.
34. Ibid.
35. Viederman, "From Prudent Man to Prudent Person," p. 8.

36. Lester R. Brown and Jennifer Mitchell, "Building a New Economy," in *State of the World 1998,* edited by Lester R. Brown (New York: W. W. Norton & Company, 1998), p. 182.
37. Richard L. Grossman, *Claiming Our Sovereignty: Establishing Control over the Corporation* (South Yarmouth, Mass.: Program on Corporations, Law & Democracy, n.d.), p. 3.
38. Robert Swann and Susan Witt, "Local Currencies," in Hannum, *People, Land, and Community.*
39. Tokar, *Earth for Sale,* p. 200.
40. Robert L. Thayer Jr., *Gray World, Green Heart* (New York: John Wiley & Sons, 1994), p. 268.

Chapter 6. Playing Politics: Environmentalists and the Electoral Process

1. Alexander Cockburn, "Clinton's Legacy Is Already in Place," *San Jose Mercury News,* 21 August 1998.
2. Michael E. Kraft, *Environmental Policy and Politics* (New York: HarperCollins, 1996), p. 192.
3. Walter A. Rosenbaum, *Environmental Politics and Policy,* 3d ed. (Washington, D.C.: CQ Press, 1995), p. 345.
4. "The Bridge to the High Road, Part 2," *Rachel's Environment & Health Weekly,* no. 619 (8 October 1998); available on-line at http://www.rachel.org/bulletin/index.cfm?St=3 (31 August 1999).
5. William Ophuls and A. Stephen Boyan Jr., *Ecology and the Politics of Scarcity Revisited: The Unraveling of the American Dream* (New York: W. H. Freeman & Company, 1992), p. 246.
6. John Rensenbrink, *The Greens and the Politics of Transformation* (San Pedro, Calif.: R. & E. Miles Publishers, 1992), pp. 38–39.
7. Will Rogers, syndicated newspaper article, 28 June 1931.
8. Figures were compiled by the Center for Responsive Politics from data released by the Federal Election Commission.
9. Max Frankel, "Save Democracy First!" *New York Times Sunday Magazine,* 21 February 1999, p. 28.
10. Allan Shuldiner and staff of the Center for Responsive Politics, "Influence, Inc.: The Bottom Line on Washington Lobbying," February 1999; available on-line at http://www.opensecrets.org/lobbyists/index.htm (25 August 1999).
11. Quoted in Sharon Beder, *Global Spin* (White River Junction, Vt.: Chelsea Green Publishing Company, 1998), p. 238.

12. See, for example, "Statement in Support of Overturning *Buckley v. Valeo,*" signed in 1997 by fifty law professors and legal experts. Distributed by USPIRG (March 1997), available on-line at http://www.pirg/demos/ch/scholars/htm
13. Ophuls and Boyan, *Ecology and the Politics of Scarcity Revisited,* pp. 243–244.
14. I encountered this timidity personally when in 1995 the Conservation Law Foundation of New England asked me to write an article for its magazine about what the 104th Congress was trying to do to roll back the environmental laws. Because I named the politicians involved, the foundation concluded that the use of such names constituted partisan activity and that therefore it could not publish the article. One lesson I learned from the episode is that some environmentalists, at least, could do with a little more political backbone.
15. Beder, *Global Spin,* p. 241.
16. Frankel, "Save Democracy First!"
17. Alison Anderson, *Media, Culture, and the Environment* (New Brunswick, N.J.: Rutgers University Press, 1997), p. 209.
18. Tom Ridge, "Needed: A G.O.P. Environmental Identity," *New York Times,* 15 March 1999, p. A25.
19. Steve Yozwiak, "Witch Hunt or Crusade?" *Arizona Republic,* 7 August 1998; available on-line at http://www.fguardians.org/news/n980807c.html (31 August 1999).
20. "GOP: Party Must Adopt Green Mantle—Sierra Club Prez," *Greenwire,* 6 December 1998; available on-line to subscribers at http://www.nationaljournal.com/aboutgreenwire.htm.
21. David Korten, *Getting to the Twenty-First Century* (West Hartford, Conn.: Kumarian Press, 1990), p. 99.
22. Quoted in Philip Shabecoff, *A Fierce Green Fire* (New York: Hill & Wang, 1993), p. 118.
23. Lynton Keith Caldwell, *Between Two Worlds: Science, the Environmental Movement, and Policy Choice* (Cambridge, England: Cambridge University Press, 1990), p. 94.
24. Mike Feinstein, "1998: Record Victories, Candidates Pace Greens," *Green Pages* (publication of the Association of State Green Parties) 3, no. 1 (winter 1999): 1.
25. "Carol Miller for Congress: New Mexico Greens Play for Keeps," *Green Pages* 1, no. 2 (fall 1997): 1.
26. "Rebuilding Democracy and Civic Spirit" (text of Ralph Nader's

acceptance speech at the Green Party's national convention), *Progressive Populist* 2, no. 10 (October 1996): 1.

27. Charlene Spretnak and Fritjof Capra, *Green Politics* (Santa Fe, N.M.: Bear & Company, 1986), pp. 231–233.
28. Willis Harman, *Global Mind Change: The Promise of the Twenty-First Century* (San Francisco: Berrett-Koehler Publishers, 1998), pp. 185–186.
29. Caldwell, *Between Two Worlds,* p. 169.
30. James MacGregor Burns, *The Deadlock of Democracy* (Englewood Cliffs, N.J.: Prentice-Hall, 1963), p. 340.

Chapter 7. Taming the Genie: Science, Technology, and Environmentalism

1. Edward O. Wilson, *Consilience* (New York: Alfred A. Knopf, 1998), pp. 266–267.
2. Ibid., p. 265.
3. John Muir, *Journal,* 27 July 1969, quoted in *A Dictionary of Environmental Quotations,* edited by Barbara K. Rodes and Rice Odell (New York: Simon & Schuster, 1992), p. 56.
4. Francis Bacon, "The Goal of Science," quoted in *A Documentary History of Conservation in America,* edited by Robert McHenry and Charles Van Doren (New York: Praeger Publishers, 1972), p. 104.
5. Francis Bacon, "The New Atlantis," in *The Harvard Classics,* edited by Charles W. Eliot (New York: P. F. Collier & Son Company, 1909), p. 181.
6. Theodore Roszak, *The Voice of the Earth* (New York: Simon & Schuster, 1992), p. 256.
7. Silvio O. Funtowicz and Jerome R. Ravetz, "A New Scientific Methodology for Global Environmental Issues," in *Ecological Economics,* edited by Robert Costanza (New York: Columbia University Press, 1991), p. 138.
8. Willis Harman, *Global Mind Change: The Promise of the Twenty-First Century* (San Francisco: Berrett-Koehler Publishers, 1998), p. 116.
9. Robert L. Thayer Jr., *Gray World, Green Heart* (New York: John Wiley & Sons, 1994), p. 55.
10. Stephen F. Mason, *A History of the Sciences* (New York: Collier Books, 1962), p. 602.

11. Lester W. Milbrath, *Envisioning a Sustainable Society: Learning Our Way Out* (Albany: State University of New York Press, 1989), p. 253.
12. Available on-line at http://www.ucsusa.org/resources/warning.html (29 August 1999).
13. George Schaller, speech given at conference on Nature and Human Society, Library of Congress, Washington, D.C., 28 October 1997.
14. Jane Lubchenco, “Entering the Century of the Environment: A New Social Contract for Science,” president’s address at annual meeting of the American Association for the Advancement of Science, Corvallis, Oregon, 15 February 1997 (text reprinted in *Science,* 23 January 1998; available on-line to subscribers at http://www.sciencemag.org/).
15. Ibid.
16. “PCBs: GE-Funded Study Finds No Occupational Cancer Risk,” *Greenwire,* 10 March 1999; available on-line to subscribers at http://www.nationaljournal.com/aboutgreenwire.htm.
17. Lynton Keith Caldwell, *Between Two Worlds: Science, the Environmental Movement, and Policy Choice* (Cambridge, England: Cambridge University Press, 1992), p. 14.
18. Ralph Nader, “Real Junk Science: The Corruption of Science by Corporate Money,” *New Solutions* 8, no. 1 (1998): 44.
19. Carl Pope, “Science at War with Itself,” *Sierra* (March–April 1998): 18.
20. Nader, “Real Junk Science,” p. 34.
21. Paul Ehrlich, “Call It Brownlash,” *World Watch* (October 1995): 5.
22. Milbrath, *Envisioning a Sustainable Society,* p. 253.
23. Ibid.
24. “The Heidelberg Appeal” (unpaginated Internet download; source not identified).
25. Milbrath, *Envisioning a Sustainable Society,* p. 252.
26. Ibid., pp. 249 ff.
27. Kai N. Lee, *Compass and Gyroscope: Integrating Science and Politics for the Environment* (Washington, D.C.: Island Press, 1993), p. 161.
28. Funtowicz and Ravetz, “A New Scientific Methodology,” p. 151.
29. Ibid., pp. 14 ff.
30. Carroll L. Bastian, ed., *Toxics Watch 1995* (New York: INFORM, 1995), p. 76.
31. Charles Perrings, “Reserved Rationality and the Precautionary Principle: Technological Change, Time, and Uncertainty in Environmental Decision Making,” in Costanza, *Ecological Economics,* pp. 191–192.

32. Funtowicz and Ravetz, "A New Scientific Methodology," p. 150.
33. Devra Davis, quoted in Philip Shabecoff, *A Fierce Green Fire* (New York: Hill & Wang, 1993), p. 135.
34. Caldwell, *Between Two Worlds*, p. 60.

Chapter 8. Small World: America and the Global Environment

1. For a fuller account of the Earth Summit, international environmentalism, and green geopolitics, see, among other sources, Philip Shabecoff, *A New Name for Peace: International Environmentalism, Sustainable Development, and Democracy* (Hanover, N.H.: University Press of New England, 1996).
2. Thomas L. Friedman, "A Manifesto for the Fast World," *New York Times Sunday Magazine*, 28 March 1999, p. 42.
3. Paul Lewis, "As Nations Shed Roles, Is Medieval the Future?" *New York Times*, 2 January 1999, p. A15.
4. Data on global environmental and economic trends are from Lester R. Brown, Christopher Flavin, and Hilary French, eds., *State of the World 1998* (New York: W. W. Norton & Company, 1998).
5. Henry W. Kendall et al., *Meeting the Challenges of Population, Environment, and Resources: The Costs of Inaction*, Environmentally Sustainable Development Proceedings Series, no. 14 (Washington, D.C.: World Bank, 1996), p. 38.
6. Commission on Global Governance, *Our Global Neighborhood* (Oxford, England: Oxford University Press, 1995), p. xix.
7. Benjamin R. Barber, *Jihad vs. McWorld* (New York: Ballantine Books, 1996), p. 85.
8. Ibid., p. 149.
9. "Global Finance Crisis: The Eye of the Storm" (Siena Declaration on the Crisis of Economic Globalization), advertisement by the International Forum on Globalization, *New York Times*, 20 November 1998, p. C7.
10. Russell Mokhiber, "Daishowa, Inc.: Kill the Messenger," *Multinational Monitor* 17, no. 12 (December 1996); available on-line at http://www.ratical.com/corporations/mm10worst96.html#n3 (24 August 1999).
11. "Green Business: Firms Must Take Broader View" (summary of article by Stuart Hart in the 2 January 1997 issue of the *Harvard Business Review*), *Greenwire*, 8 January 1997; available on-line to subscribers at http://www.nationaljournal.com/aboutgreenwire.htm.

12. Quoted in *Business Week,* 10 November 1997, as summarized in *Greenwire,* 7 November 1997; available on-line to subscribers at http://www.nationaljournal.com/aboutgreenwire.htm.
13. Tim Shorrock, "Asian Financial Crisis," *Corporate Watch* 3, no. 8 (April 1998); available on-line at http://www.corpwatch.org/trac/globalization/financial/focus.html (31 August 1999).
14. *Caring for the Earth: A Strategy for Sustainable Living* (Gland, Switzerland: IUCN—The World Conservation Union, United Nations Environment Programme, and World Wide Fund for Nature, 1991).
15. Kai N. Lee, *Compass and Gyroscope: Integrating Science and Politics for the Environment* (Washington, D.C.: Island Press, 1993), p. 9.
16. Martin Holdgate, *From Care to Action: Making a Sustainable World* (London: Earthscan Publications, 1996), p. 116.
17. Ann Swardson, "Turtle Protection Law Overturned by WTO, Environmentalists Angered by Decision," *Washington Post,* 13 October 1998, p. C2.
18. Durwood Zaelke, Paul Orbuch, and Robert F. Housman, eds., *Trade and the Environment: Law, Economics, and Policy* (Washington, D.C.: Island Press, 1993), p. xiv.
19. Elizabeth Dowdeswell and Steve Charnovitz, "Globalization, Trade, and Interdependence," in *Thinking Ecologically,* edited by Marion R. Chertow and Daniel C. Esty (New Haven, Conn.: Yale University Press, 1997), p. 93.
20. Ibid., p. 101.
21. Gareth Porter and Janet Welsh Brown, *Global Environmental Politics* (Boulder, Colo.: Westview Press, 1991), p. 153.
22. Lynton Keith Caldwell, *International Environmental Policy,* 2d ed. (Durham, N.C.: Duke University Press, 1990), pp. 304–305.
23. Commission on Global Governance, *Our Global Neighborhood,* p. 227.
24. Conor Cruise O'Brien, essay in *New Republic,* 4 November 1986, quoted in *The Columbia Dictionary of Quotations* (New York: Columbia University Press, 1993) (unpaginated Internet download).
25. Daniel Patrick Moynihan, *On the Laws of Nations* (Cambridge, Mass.: Harvard University Press, 1990), p. 120.
26. Ibid., p. 177.
27. Barber, *Jihad vs. McWorld,* p. 12.
28. Commission on Global Governance, *Our Global Neighborhood,* p. 2.
29. Porter and Brown, *Global Environmental Politics,* p. 152.
30. Sir Crispin Tickell, "Grasping the Concept of Environmental Secu-

rity," in *Threats without Enemies,* edited by Gwyn Pryns (London: Earthscan Publications, 1993), p. 23.

31. Commission on Global Governance, *Our Global Neighborhood,* p. 335.
32. Admiral Sir Julian Oswald, "Accepting the Challenge of Environmental Security," in Pryns, *Threats without Enemies,* pp. 115–118.
33. Thomas F. Homer-Dixon, *Environmental Change and Violent Conflict,* Emerging Issues Occasional Paper Series, no. 4 (Cambridge, Mass.: American Academy of Arts and Sciences, June 1990), p. 2.
34. Johan Holmberg and Richard Sandbrook, "Sustainable Development: What Is to Be Done?" in *Making Development Sustainable: Redefining Institutions, Policy, and Economics,* edited by Johan Holmberg (Washington, D.C.: Island Press, 1992).
35. Riley E. Dunlap, "International Opinion at the Century's End: Public Attitudes toward Environmental Issues," in *Environmental Policy: Transnational Issues and National Trends,* edited by Lynton Keith Caldwell and Robert V. Bartlett (Westport, Conn.: Quorum Books, 1997), p. 201.

Chapter 9. Transforming the Future

1. Rachel Carson, *Silent Spring,* 25th anniversary ed. (Boston: Houghton Mifflin Company, 1987), p. 279.
2. Paul Raskin et al., "Bending the Curve: Toward Global Sustainability," report of the Global Scenario Group, Stockholm Environment Institute, 1998; summary available on-line at http://www.gsg.org/btcsum.html (31 August 1999).
3. *Tax Waste, Not Work,* Executive Summary (San Francisco: Redefining Progress, April 1997), p. 12; available on-line at http://www.rprogress.org/pubs/twnw/twnw_execsum.html (24 August 1999).
4. Michael E. Kraft, *Environmental Policy and Politics* (New York: HarperCollins, 1996), pp. 154–157.
5. John Rensenbrink, *Against All Odds: The Green Transformation of American Politics* (Raymond, Maine: Leopold Press, 1999), p. 13.
6. Ibid., p. 228.

Interviews

The men and women on this list were interviewed by the author during the years 1998 and 1999. All of the direct and indirect quotations from them in the text are from those interviews, unless otherwise specified.

John Adams
Jonathan Adler
Doug Bailey
Spencer Beebe
Brent Blackwelder
Sherwood Boehlert
David Buzzelli
Deb Callahan
Eric Chivian
Philip Clapp
William Clark
Barry Commoner
Anthony Cortese
Douglas Costle
Herman Daly
J. Clarence (Terry) Davies
Barbara Dudley
Riley Dunlap
David Ehrenfeld
Robert Engelman
Daniel C. Esty
Michael Fischer
John Flicker
Michael Frome
Kathryn Fuller
Lois Gibbs
Paul Gorman
Wade Greene
Karl Grossman
Richard Grossman
David Hahn-Baker
Paul Hawken
David Hawkins
Tom Hayden
Denis Hayes
Samuel Hays
Tim Hermach
Allen Hershkowitz
Jean Hocker
Philip Hocker
Huey Johnson
Dale Jorgenson
Thomas Jorling
Gene Karpinski
Daniel Kemmis
Henry Kendall
Kalee Kreider
Fred Krupp
Laurie Lane-Zucker
Jonathan Lash
Lester Lave
Charles Little

Michael McCloskey
William Meadows
Lester Milbrath
Curtis Moore
Richard Moore
Richard Morgenstern
Gaylord Nelson
David Orr
Patrick Parenteau
John Passacantando
Jane Perkins
Russell Peterson
Bruce Piasecki
Carl Pope
Paul Pritchard
Peter Raven
Joshua Reichert
William Reilly
John Rensenbrink
Arlie Schardt
Robert Shapiro
Peggy Shepard
William Shutkin
Charlene Spretnak
Maurice Strong
Terry Swearingen
Dan Swinney
Lee Thomas
Joanna Underwood
Mark Van Putten
Stephen Viederman
Norman Vig
Konrad von Moltke
Morris (Bud) Ward
Susan Witt
George Woodwell

Bibliography

Adler, Jonathan. *Environmentalism at the Crossroads: Green Activism in America*. Washington, D.C.: Capital Research Center, 1995.

Anderson, Alison. *Media, Culture, and the Environment*. New Brunswick, N.J.: Rutgers University Press, 1997.

Ayers, Harvard, Jenny Hager, and Charles E. Little, eds. *An Appalachian Tragedy: Air Pollution and Tree Death in the Highland Forest of Eastern North America*. San Francisco: Sierra Club Books, 1998.

Barber, Benjamin R. *Jihad vs. McWorld*. New York: Ballantine Books, 1996.

Bastian, Carroll L., ed. *Toxics Watch 1995*. New York: INFORM, 1995.

Beder, Sharon. *Global Spin*. White River Junction, Vt.: Chelsea Green Publishing Company, 1998.

Bellah, Robert N., Richard Madsen, William M. Sullivan, Ann Swidler, and Steven M. Tipton. *Habits of the Heart*. New York: Harper & Row, 1985.

Brown, Lester R., Christopher Flavin, and Hilary French, eds. *State of the World 1998*. New York: W. W. Norton & Company, 1998.

Brown, Lester R., Michael Renner, and Christopher Flavin. *Vital Signs 1997*. New York: W. W. Norton & Company, 1997.

Bullard, Robert D., ed. *Unequal Protection*. San Francisco: Sierra Club Books, 1994.

Burns, James MacGregor. *The Deadlock of Democracy*. Englewood Cliffs, N.J.: Prentice-Hall, 1963.

Caldwell, Lynton Keith. *Between Two Worlds: Science, the Environmental Movement, and Policy Choice*. Cambridge, England: Cambridge University Press, 1992.

———. *International Environmental Policy*. 2d ed. Durham, N.C.: Duke University Press, 1990.

Caldwell, Lynton Keith, and Robert V. Bartlett, eds. *Environmental Policy: Transnational Issues and National Trends*. Westport, Conn.: Quorum Books, 1997.

Carson, Rachel. *Silent Spring*. 25th anniversary ed. Boston: Houghton Mifflin Company, 1987.

Chertow, Marion R., and Daniel C. Esty, eds. *Thinking Ecologically*. New Haven, Conn.: Yale University Press, 1997.

Cisneros, Henry G., ed. *Interwoven Destinies: Cities and the Nation*. New York: W. W. Norton & Company, 1993.

Cohen, Joel E. *How Many Humans Can the Earth Support?* New York: W. W. Norton & Company, 1995.

Commission on Global Governance. *Our Global Neighborhood*. Oxford, England: Oxford University Press, 1995.

Costanza, Robert, ed. *Ecological Economics*. New York: Columbia University Press, 1991.

Daly, Herman E., and Kenneth N. Townsend. *Valuing the Earth*. Cambridge, Mass.: MIT Press, 1993.

Dowie, Mark. *Losing Ground*. Cambridge, Mass.: MIT Press, 1995.

Enterprise for the Environment. *The Environmental Protection System in Transition*. Washington, D.C.: CSIS Press, 1997.

Frome, Michael. *Green Ink*. Salt Lake City: University of Utah Press, 1998.

Gottlieb, Robert. *Forcing the Spring: The Transformation of the American Environmental Movement*. Washington, D.C.: Island Press, 1993.

Grossman, Richard L., and Frank T. Adams. *Taking Care of Business: Citizenship and the Charter of Incorporation*. Cambridge, Mass.: Charter, Ink, 1993.

Hannum, Hildegarde, ed. *People, Land, and Community*. New Haven, Conn.: Yale University Press, 1997.

Harman, Willis. *Global Mind Change: The Promise of the Twenty-First Century*. San Francisco: Berrett-Koehler Publishers, 1998.

Hawken, Paul. *The Ecology of Commerce*. New York: HarperCollins, 1993.

Hayden, Tom. *The Lost Gospel of the Earth: A Call for Renewing Nature, Spirit, and Politics*. San Francisco: Sierra Club Books, 1996.

Hays, Samuel B. *Beauty, Health, and Permanence*. Cambridge, England: Cambridge University Press, 1987.

Henderson, Hazel. *Building a Win-Win World*. San Francisco: Berrett-Koehler Publishers, 1996.

Holdgate, Martin. *From Care to Action: Making a Sustainable World*. London: Earthscan Publications, 1996.

Holmberg, Johan, ed. *Making Development Sustainable: Redefining Institutions, Policy, and Economics*. Washington, D.C.: Island Press, 1992.

Homer-Dixon, Thomas F. *Environmental Change and Violent Conflict*. Emerging Issues Occasional Paper Series, no. 4. Cambridge, Mass.: American Academy of Arts and Sciences, June 1990.

Kemmis, Daniel. *Community and the Politics of Place.* Norman: University of Oklahoma Press, 1990.

Korten, David. *Getting to the Twenty-First Century.* West Hartford, Conn.: Kumarian Press, 1990.

Kraft, Michael E. *Environmental Policy and Politics.* New York: HarperCollins, 1996.

Lee, Kai N. *Compass and Gyroscope: Integrating Science and Politics for the Environment.* Washington, D.C.: Island Press, 1993.

Leopold, Aldo. *A Sand County Almanac.* New York: Oxford University Press, 1948.

McHenry, Robert, and Charles Van Doren, eds. *A Documentary History of Conservation in America.* New York: Praeger Publishers, 1972.

Marsh, George Perkins. *Man and Nature.* Edited by David Lowenthal. 1864. Reprint, Cambridge, Mass.: Harvard University Press, Belknap Press, 1965.

Mason, Stephen F. *A History of the Sciences.* New York: Collier Books, 1962.

Milbrath, Lester W. *Envisioning a Sustainable Society: Learning Our Way Out.* Albany: State University of New York Press, 1989.

Moynihan, Daniel Patrick. *On the Laws of Nations.* Cambridge, Mass.: Harvard University Press, 1990.

Ophuls, William, and A. Stephen Boyan Jr. *Ecology and the Politics of Scarcity Revisited.* New York: W. H. Freeman & Company, 1992.

Orr, David W. *Earth in Mind: On Education, Environment, and the Human Prospect.* Washington, D.C.: Island Press, 1994.

———. *Ecological Literacy.* Albany: State University of New York Press, 1992.

Papanek, Victor. *The Green Imperative.* New York: Thames & Hudson, 1995.

Pigou, A. C. *Economics of Welfare.* 1920. Reprint, London: Macmillan, 1952.

Porter, Gareth, and Janet Welsh Brown. *Global Environmental Politics.* Boulder, Colo.: Westview Press, 1991.

President's Council on Sustainable Development. *Sustainable America: A New Consensus for Prosperity, Opportunity, and a Healthy Environment for the Future.* Washington, D.C.: President's Council on Sustainable Development, 1996.

Pryns, Gwyn, ed. *Threats without Enemies.* London: Earthscan Publications, 1993.

Rensenbrink, John. *Against All Odds: The Green Transformation of American Politics.* Raymond, Maine: Leopold Press, 1999.

———. *The Greens and the Politics of Transformation*. San Pedro, Calif.: R. & E. Miles Publishers, 1992.

Rodes, Barbara K., and Rice Odell, eds. *A Dictionary of Environmental Quotations*. New York: Simon & Schuster, 1992.

Rosenbaum, Walter A. *Environmental Politics and Policy*. 3d ed. Washington, D.C.: CQ Press, 1995.

Roszak, Theodore. *The Voice of the Earth*. New York: Simon & Schuster, 1992.

Schumacher, E. F. *Small Is Beautiful*. New York: Harper & Row, 1973.

Shabecoff, Philip. *A Fierce Green Fire*. New York: Hill & Wang, 1993.

———. *A New Name for Peace: International Environmentalism, Sustainable Development, and Democracy*. Hanover, N.H.: University Press of New England, 1996.

Silver, Cheryl Simon, with Ruth S. DeFries. *One Earth, One Future: Our Changing Global Environment*. Washington, D.C.: National Academy Press, 1990.

Spretnak, Charlene, and Fritjof Capra. *Green Politics*. Santa Fe, N.M.: Bear & Company, 1986.

Thayer, Robert L., Jr. *Gray World, Green Heart*. New York: John Wiley & Sons, 1994.

Tokar, Brian. *Earth for Sale*. Boston: South End Press, 1997.

Vig, Norman J., and Michael E. Kraft, eds. *Environmental Policy in the 1990s*. Washington, D.C.: CQ Press, 1997.

Werbach, Adam. *Act Now, Apologize Later*. New York: HarperCollins, 1997.

Wilson, Edward O. *Consilience*. New York: Alfred A. Knopf, 1998.

World Commission on Environment and Development. *Our Common Future*. New York: Oxford University Press, 1987.

Index

Adams, John, 186
Adaptive management, 188
Adler, Jonathan, 85
AFL-CIO, 67, 68
Agenda 21, 156, 162
Agriculture, 20, 21
  community-supported, 76–78, 109
  synthetic chemicals used in, 4–5, 16
Air pollution, 4, 17, 18, 21, 87
  tradable permits, 9, 88–89, 187
Alaska National Interest Lands Conservation Act, 128
Alliance for Environmental Innovation, The, 105
Alternatives for Community & Environment, 56–57, 58, 132
American Coal Foundation, 70
American Forests, 34
Amir, Sara, 133
Anderson, Alison, 121–22
Anderson, Ray, 100
Arcata, California, 106
Arrow, Kenneth, 93
Asian financial crisis, 160
Augustine, Rose, 55–56

Bacon, Francis, 138, 139
Bailey, Doug, 113, 116, 120
Banana Kelly, 62, 65, 107–108
Barber, Benjamin R., 158–59, 169
Bastian, Ann, 69
Beder, Sharon, 120
Beebe, Spencer, 35, 55
Bell, Daniel, 83
Bellah, Robert, 57, 59, 81
Benedick, Richard E., 148
Berdes, John, 63
Berkshire County, Massachusetts, 76–78, 109
Berry, Thomas, 28, 74–75
Biological diversity, 19, 145
Blackwelder, Brent, 44, 172
Boehlert, Sherwood, 35, 123
Boise Cascade Corporation, 69
Bookchin, Murray, 59
Bosso, Christopher J., 37
Boyan, A. Stephen, Jr., 27, 98, 115, 118–19
Brandt, Willy, 169
British Petroleum Company, 98
Bronx Community Paper Company, 62, 63, 65, 107–108, 182–83
Brower, David, 186
Brown, Janet Welsh, 167
Browne, John, 98
Brownfields, 86
Buckley v. Valeo, 117–18
Bullard, Robert D., 64
Bulletin of the Atomic Scientists, 147
Burns, James McGregor, 136
Bush, George, 8
Buzzelli, David, 101, 160

Caldwell, Lynton Keith, 31, 136, 143
Callahan, Deb, 45, 113, 116, 119, 126
Campaign finance reform, 117–18
Capitalism, *see* Market capitalism
Caring for the Earth, 162
Carson, Rachel, 5, 16, 140, 177, 178
Center for Health, Environment and Justice, 35–36, 44, 45, 68–69, 128, 130, 181

CFCs (chlorofluorocarbons), 151–52
Chafee, John, 123
Charnovitz, Steve, 165, 166
Chemical Manufacturers Association, 102
China, 161–62
Chivian, Dr. Eric, 153
"Civic science," 147
Clapp, Philip, 38, 113, 123, 132
Clark, William, 17, 30, 46, 118
Clean Air Act of 1970, 6, 78
  Amendments of 1990, 9, 88–89
Climate change, *see* Global warming
Clinton, Bill, 36, 84–85, 111, 112, 119–20, 169
Cockburn, Alexander, 112
Commission on Global Governance, 157, 167–68, 169–70
Committee for the National Institute for the Environment (CNIE), 148
Commoner, Barry, 5, 22, 140
Communism, 83, 84, 87
Community and environmentalism, 34–35, 53–63, 78–81, 181
  community-based development, 59–63, 109, 180–81, 182–83, 185
Community and the Politics of Place (Kemmis), 78
Competitive Enterprise Institute, 85
Congress, *see* U.S. Congress
Consilience (Wilson), 137
"Contract with America," 88, 123
Corporate America, 70, 95–102, 187
  brownlash by, 144
  competing instincts of, 101
  environmentalists and, 103–10
  global economy and, *see* Global economy
  "greening" of, 22–23, 98, 183
  legal standing of corporations, 97, 106–107
  lobbying by, 96, 102, 116
  Natural Step approach and, 99
  resistance to environmentalism, 8, 23, 50–51
Corporate charters, 106
Cortes, Ernesto, Jr., 65–66
Cortese, Anthony, 35, 70–71, 178
Costanza, Robert, 87

Daishowa Inc., 159
Daly, Herman, 86, 94, 102, 162
Davis, J. Clarence (Terry), 22–23
Deadlock of Democracy, The (Burns), 136
Defenders of Wildlife, 3
Deforestation, 19
DeLay, Tom, 124, 143–44
Democracy, 138
  environmentalism and, 7
  market capitalism and, 83
  power of corporations and, 96
  voter participation, 114–15, 135
Democratic Party, 115, 124–25, 134
Devall, Bill, 33–34
Dow Chemical Company, 101
Dowdeswell, Elizabeth, 165, 166
Dowie, Mark, 10
Ducks Unlimited, 3
Dudley, Barbara, 30, 44
Dunlap, Riley E., 25, 31, 50–51, 175
du Pont de Nemours and Company, E. I., 102

Earth Day, 5, 23, 186
Earth First!, 130
Earth Summit of 1992, Rio de Janeiro, 24, 94, 145, 155–56, 167, 171
Economy and environmentalism, 83–110
  corporate America, *see* Corporate America
  economic growth, 89–95, 161–62
  environmentalists and, 103–10
  global economy, *see* Global economy
  market capitalism, *see* Market capitalism
  transforming the economy, 186–89
Ecotrust, 34–35, 55, 58, 63, 188–89
Education and environmentalism, 69–71, 124

Ehrenfeld, David, 17
Ehrlich, Anne, 140
Ehrlich, Paul, 5, 140, 144
Electoral process, environmental and, *see* Politics, environmental
Electric Power Research Institute, 30
End of Ideology, The (Bell), 83
Enlibra doctrine, 80–81
Enterprise for the Environment, 22
Environmental Action, 6–7
Environmental audits, 103
Environmental Defense Fund, 6, 9, 18, 29, 33, 40–41, 171
Environmental Grantmakers Association, 39
Environmental justice movement, 34, 47–48, 63–66, 132, 180
Environmental Media Services, 37, 72, 117, 184
Environmental movement, 1–11, 16–17
  allies, need for, 62, 66, 68–69, 128–33, 180–81, 190–91
  current state of, 29–52
  defining the, 30–36
  divisions in, 33–34
  education and, 69–71, 124
  first wave of, 1–3
  foundations and, 38–40, 182
  fourth wave of, 185
  fund-raising by, 181–82, 183
  future of, 177–94
  global environment and, 155–75
  leadership of, 45–46, 50, 129, 186
  media and, *see* Media coverage of environmental issues
  organized labor and, 66–69
  racism and, *see* Environmental justice movement
  Reagan administration and, *see* Reagan administration
  second wave of, 5–7
  strengths of, 49
  third wave of, 8–9, 88
  values of, 192
  weaknesses of, 50–52
Environmental Policy Center, 34
Environmental Protection Agency (EPA), 6, 8, 23, 66, 122, 149
  Toxics Release Inventory (TRI), 103, 149
Environmental Working Group, 34
Esty, Daniel C., 70, 114, 165, 188
Ethnic, tribal, and religious conflicts, 26, 155
European Union, 164, 168
Exxon Corporation, 70

Farming, *see* Agriculture
Federal Trade Commission, 149
Federal Water Pollution Control Act Amendments of 1972, 6
Ferris, Deeohn, 63
First National Bank of Boston v. Bellotti, 106
First National People of Color Environmental Leadership Summit, 64
Fischer, Michael, 61
Flicker, John, 41, 135
Food and Drug Administration, 149, 151
Ford, Gerald, 147
Fossil fuels, 16, 18, 20
Foundations, 38–40, 182
  see also individual foundations
Freedom Forum's Newseum, 73
Friedman, Thomas L., 156
Friends of the Earth, 6–7, 33, 44, 103
Frome, Michael, 45–46, 76
Fuller, Kathryn, 50, 172–73
Funtowicz, Silvio O., 148
Future of environmental movement, 177–94

Gasoline prices, 86–87, 188
General Agreement of Tariffs and Trade (GATT), 165, 166
General Electric Company, 101, 142, 143
Genetic engineering, 21, 151
Gibbs, Louis, 7, 36, 44, 68–69, 126, 128, 130, 181
Gingrich, Newt, 123

Global economy, 15–16, 22, 97, 156, 158–63, 172
inequities of, 15–16, 85, 157
international trade, 163–66
Global environment, United States and, 155–75
Global Environment Facility, 191
Global warming, 18–19, 29–30, 67, 70, 91–92, 98, 143, 144, 145, 170–71
Gore, Al, 36, 94, 112, 116
Gorman, Paul, 75–76
Gorsuch, Anne Buford, 8, 122, 123
Government:
-funded science, 144–45
subsidies, 104
U.S. Congress, *see* U.S. Congress
Grass-roots organizations, 35–36, 44–45, 59, 61, 72, 174–75, 181, 182
Gray, John, 84
Great Lakes United, 34, 45
Greene, Wade, 38
Green parties, 133–35, 189, 190, 191
Greenpeace, 6–7, 43–44, 143
Grim, John, 74
Grossman, Karl, 133, 134, 173
Grossman, Richard L., 106–107

Habits of the Heart (Bellah et al.), 57, 59, 81
Hague Declaration, 170
Hahn-Baker, David, 46–47, 61, 192
Hamburg, Dan, 133
Hardin, Garrett, 5
Harman, William, 134, 139
Hart, Stuart, 160
Harvard Business Review, 160
Hawken, Paul, 100, 160
Hawkins, David, 129–30
Hayden, Tom, 46, 75, 117, 119
Hayes, Denis, 171, 186
Hays, Samuel P., 3
Hazardous waste, 4, 64, 90, 103
"Heidelberg Appeal," 145–47
Henderson Hazel, 86
Herger, Wally, 79–80
Heritage Foundation, 40
Hermach, Tim, 192
Hershkowitz, Allen, 108, 183, 187
Hocker, Jean, 43
Hocker, Philip, 192
Holdgate, Sir Martin, 162
Holism, 137–38, 142, 179
Homer-Dixon, Thomas F., 173–74
Hotline, 113
Housman, Robert, 163
Hubbard Brook Experimental Forest, 18
Hughes Aircraft Company, 55–56

INFORM, Inc., 102, 103
Inglis, David, 77
Innovest Strategic Value Advisors, Inc., 102
International Forum on Globalization, 159
International Institute for Sustainable Development, 174–75
Internet, 149, 185
ISO 14000, 103
Izaak Walton League of America, 3

Jackson, Dana Lee, 57
Jesse Smith Noyes Foundation, 39, 129
Job Creation and Enhancement Act, 123
Job layoffs, environmental protection and, 90, 92
Jorgenson, Dale, 92–93, 105, 159, 165
Jorling, Thomas C., 46, 88–89
Journal for Occupational and Environmental Medicine, 142

Karpinski, Gene, 117, 171
Kemmis, Daniel, 79, 80, 127
Kendall, Henry W., 20, 152, 157
Kiernan, Matthew J., 102
Kitzhaber, John, 81
Korten, David, 21, 50, 62, 127
Kraft, Michael, 10, 27, 45, 112, 123, 189
Kreider, Kalee, 43–44, 50, 129

Krimsky, Sheldon, 147
Krupp, Fred, 29, 40–41, 88, 170, 185
Kuttner, Robert, 87
Kyoto treaty on global warming, 29–30, 67, 98, 170–71

Labor, organized, 66–69, 132, 181
Land Institute, 57
Land Trust Alliance, 43
Lash, Jonathan, 39, 162
Lave, Lester, 22
League of Conservation Voters, 7, 45, 113, 116, 119, 124, 126, 190
Lee, Kai N., 147–48, 162
Legislation, environmental:
  international, 173
  of 1970s, 5–6, 23, 85, 130
Leopold, Aldo, 5, 51, 56
Lewis, John, 65
Likens, Gene, 18
Little, Charles, 19, 31, 192
Lobbying, 116
  by corporations, 96, 102, 116
  by environmental groups, 41, 118–19
Long Island Lighting Company, 89
Los Angeles Times, 75
Love Canal, 7
Lubchenco, Jane, 13, 141–42, 146, 153

MacArthur Foundation, John D. and Catherine, 39
McCloskey, Michael, 33, 51, 57, 80, 113, 186
McGrady, Chuck, 125
Man and Nature (Marsh), 2
Marine species, threat to, 19–20
Market capitalism, 83–84, 109, 187
  environment and, 84–89
Marsh, George Perkins, 2
Mason, Stephen F., 140
MAXXAM Corporation, 106
Meadows, William, 43
Media coverage of environmental issues, 27, 71–74, 96, 119–22, 184–85
Mertig, Angela G., 31
Mexico, 17, 164, 174
Midwest Center for Labor Research, 107
Milbrath, Lester, 27, 31, 70, 140, 144, 145, 147
Miller, Carol, 133
Minow, Newt, 120
Molina, Mario, 151
Monsanto Company, 99–100, 143, 160
Moore, Richard, 48, 61, 174
Morrison, Denton, 129
Moynihan, Daniel Patrick, 169
Muir, John, 1, 3, 11, 51, 137–38
Muskie, Edmund, 122

Nader, Ralph, 33, 133, 134, 143
National Academy of Public Administration, 89
National Academy of Sciences, 14
National Audubon Society, 3, 7, 41
National Environmental Policy Act, 6
National Environmental Trust, 34, 37, 38, 72, 113, 184
National Forest System, 3
National Institute for the Environment, proposed, 148, 153
National Library for the Environment, proposed, 148
National Parks and Conservation Association, 3
National Religious Partnership for the Environment, 75
National Safety Council, Environmental Health Care Center, 66
National Wildlife Federation, 3, 7, 33, 35, 42, 171
National Wildlife Refuge System, 3
Native Forest Council, 29
Natural Resources Defense Council, 6, 38–39, 41, 53–54, 58, 117, 129, 171, 186
  Bronx Community Paper Company and, 62, 63, 65, 107–108, 182–83
Nature Conservancy, 43, 55

Nelson, Gaylord, 58, 124, 186
"New Atlantic, The," 138
"New Scientific Methodology for Global Environmental Issues, A," 148
New York Times, 123, 159
Nixon, Richard, 5, 122
North American Free Trade Agreement (NAFTA), 26, 33, 163, 164–65, 191

O'Brien, Conor, 168
Occupational Safety and Health Act, 6
Occupational Safety and Health Administration, 66
Open space, 43, 77
Ophuls, William, 27, 98, 115, 118–19
Oppenheimer, Michael, 170
Orbuch, Paul, 163
Orr, David, 26–27, 70, 71
Osborn, Fairfield, 5
Oswald, Sir Julian, 173
Our Common Future, 94
Ozone layer, destruction of, 20, 151–52

Papanek, Victor, 98
Parenteu, Patrick, 89, 192
Passacantando, John, 129
Pataki, George, 89, 123
PCBs (polychlorinated biphenyls), 101, 142
People-Centered Development Forum, 21
Perkins, Jane, 67–68
Perrings, Charles, 150
Peterson, Russell, 124
Pew Charitable Trusts, 10, 38, 39, 184
Physicians for Social Responsibility, 7, 132, 140
Piaseckik, Bruce, 101
Pigou, A. C., 86
Pimentel, David, 20
Pinchot, Gifford, 1, 3, 11
Place-based conservation, 43, 56, 76–81, 127
Politics, environmental, 23–25, 36–37, 96–97, 111–36, 184, 189–91
  forming alliances, *see* Environmental movement, allies, need for
  Green parties, 133–35, 189, 190, 191
  money and, 115–19
  organizing for power, 126–28
  see also U.S. Congress
Pollution, 4, 6, 18, 20
  air, *see* Air pollution
Pope, Carl, 42, 168
Population growth, 14–15
Porter, Gareth, 167
Precautionary principle, 149–51
President's Council on Sustainable Development, 61–62, 101
Property rights groups, 40, 79, 114
PR Watch, 116

Quincy, California, 79–80

Raven, Peter, 14, 16, 58, 98, 140
Ravetz, Jerome R., 148
Reagan administration, 7–8, 23–24, 88, 112, 122–23
Redefining Progress, 188
Regional control of environmental issues, 34, 45, 79–80, 127–28
Reichert, Joshua, 10, 38, 39, 171
Reilly, William K., 46, 170
Religion and the environment, 74–76
Rensenbrink, John, 115, 134–35, 189
Republican Party:
  anti-environmental agenda of, 24, 36–38, 88, 111–12, 114, 115, 122–25
  environmental advocates in, 122, 123
Resources for the Future, 91
Richardson, Bill, 133
Ridge, Tom, 123
Right-to-know laws, 149
Rio de Janeiro, Earth Summit of 1992 in, 24, 94, 145, 155–56, 167, 171
Robèrt, Karl-Henrik, 99
Rocky Mountain Institute, 92

Roosevelt, Theodore, 1, 3, 11
Roper Center for Public Opinion Research, 73
Rosenbaum, Walter, 26, 31, 90, 112
Roszak, Theodore, 138
Rowland, Sherwood, 151
Ruckelshaus, William D., 22

Sand Country Almanac (Leopold), 5
San Francisco Chronicle, 80
Santa Clara County v. Southern Pacific, 97
Schaller, George, 19, 141
Schardt, Arlie, 72, 184
Schumacher, E. F., 59
Schumacher Society, E. F., 33
Science, 74
Science, technology, and environmentalism, 137–53
Second Nature, 70, 178
Selcraig, Bruce, 69–70
Shabecoff, Alice, 60
Shapiro, Robert, 99, 100, 160
SHARE, 108
Shepard, Peggy, 53–54
Shorebank Corporation, 55, 63
Shorebank Enterprise Pacific, 63
Shutkin, William, 39–40, 56–57, 58, 132
Sierra Club, 3, 33, 34, 42, 57, 58, 182
Silent Spring (Carson), 5, 177
Skloot, Edward, 69
SLAPP suits (strategic lawsuits against public participation), 50
Small Is Beautiful (Schumacher), 59
Smith, Adam, 84
Social contract for the scientific community, 141–42, 153
Social ecology, 59
Social justice, 54–59, 65, 190–91
  see also Environmental justice movement
Society of Environmental Journalists, 72, 121
Southwest Network for Environmental and Economic Justice, 48, 61, 174
Stafford, Robert, 122
State and local governments, environmental measures of, 6, 43
Stauber, John, 116
Strong, Maurice, 156, 160, 174
Sunways Farm, 77
Superfund Program, 90, 102
Supreme Court, 97, 106, 117–18
Surdna Foundation, 69
Sustainable development, 94, 156, 160, 162, 173, 174–75
Swearingen, Terri, 44–45, 116
Sweeney, John, 68
Swinney, Dan, 107, 187

Tax laws, 105, 118–19, 187–88
Technology, science, and environmentalism, 137–53
Thayer, Robert L., Jr., 109–10, 139
Thoreau, Henry David, 2, 51, 194
Thurow, Lester, 85
Tickell, Sir Crispin, 170
Tokar, Brian, 59, 86
Toxic waste, 20–21
Trade, international, 163–66
Treaties, international environmental, 25
Tucker, Mary Evelyn, 74

Underwood, Joanna, 102
Union of Concerned Scientists, 7, 33, 132, 140, 152
Unions, 66–69, 132, 181
United Church of Christ, Commission for Racial Justice, 64
United Nations, 25, 26, 167–70, 173, 191
  Trusteeship Council, 169–70
United Nations Commission on Sustainable Development, 167
United Nations Conference on the Human Environment, 167
United Nations Environment Programme, 26, 162, 167, 168
United States and the global environment, 155–75
U.S. Bureau of Labor Statistics, 90

U.S. Congress, 44
anti-environmental agenda of, 24, 36–38, 88, 111–12, 114, 118, 123–24, 184
foreign aid budgets, 156
U.S. Public Interest Research Group, 33, 117, 171
Urban Environment Conference, 63–64

Van Putten, Mark, 42, 180
Viederman, Stephen, 30–31, 39, 97, 105, 129, 143, 158, 187
von Moltke, Konrad, 71–72, 171–72

Ward, Morris (Bud), 66
Warfare, 26, 155, 173–74, 191
Water, 4, 21
Watt, James, 8, 122, 123
Werbach, Adam, 34, 50
Western Governors' Association, 80
West Harlem Environmental Action, 53–54, 58, 181
Wetlands, 21
Wetstone, Greg, 117, 125, 129
White, Lynn, Jr., 74
Whitman, Christine, 123
Wilderness Society, 3, 43, 186
Willapa Alliance, 55
Wilson, Edward O., 19, 137, 140, 142
"Wise use" movement, 24–25, 40, 50, 51, 79, 114
Woodwell, George, 140, 153, 170
World Bank, 26, 172
World Commission on Environmental and Development, 94
World Conservation Union (IUCN), 162
World Resources Institute, 35, 39, 171
"World Scientists' Warning to Humanity," 140–41
World Trade Organization, 26, 163, 165–66, 168, 173, 191
Worldwatch Institute, 19, 35, 105, 171
World Wide Fund for Nature, 162
World Wildlife Fund, 171, 172

Zaelke, Durwood, 163

Island Press Board of Directors